ENVIRONMENTAL BIOLOGY

ENVIRONMENTAL BIOLOGY

Emily Rodriguez
Liam Roy
Isabella Martin

Kruger Brentt
Publishers

2023

Kruger Brentt Publishers UK. LTD.
Company Number 9728962

Regd. Office: 68 St Margarets Road, Edgware, Middlesex HA8 9UU

ISBN: 978-1-78715-092-8

Disclaimer:

Every possible effort has been made to ensure that the information contained in this book is accurate at the time of going to press, and the publisher and author cannot accept responsibility for any errors or omissions, however caused. No responsibility for loss or damage occasioned to any person acting, or refraining from action, as a result of the material in this publication can be accepted by the editor, the publisher or the author. The Publisher is not associated with any product or vendor mentioned in the book. The contents of this work are intended to further general scientific research, understanding and discussion only. Readers should consult with a specialist where appropriate.

Every effort has been made to trace the owners of copyright material used in this book, if any. The author and the publisher will be grateful for any omission brought to their notice for acknowledgement in the future editions of the book.

For information on all our publications visit our website at http://krugerbrentt.com/

PREFACE

Environmental biologists examine ecosystems with a focus on how living organisms interact with their environment. This program is relevant for today's society which is facing complex environmental issues such as conservation and environmental contamination. Every year, on 5th June, we all celebrate World Environment Day. All living beings and non-living beings present on the Earth represent the environment. Plants, creatures, water, air, and other living things exist in our environment. Our environment gets influenced by Climatic interaction, Geomorphic measures, and Hydrologic measures. The life of humans and animals is entirely dependent on climate. Our environment supports life on Earth. Everything we inhale, feel, and energy comes from the environment. The environment is considered a cover that helps sustain life on Earth. Among all the planets, it is our planet Earth that supports life. Every day we get to hear about threats to the environment. Our environment includes everything from the forests to the oceans, which impacts our everyday life. It can be deforestation, pollution, soil erosion, etc., which needs to be addressed seriously. Billions of people depend on the environment for their livelihood. For example, over 1.5 billion people depend on forests for food, medicine, shelter and more. Farmers turn to the woods when their crops fail. Almost two billion people earn a living from agriculture, and the other three billion people are on the ocean. Many negative consequences are encountered due to biodiversity loss, but weakened food security is extensive. If we lose our precious animals and plant species, we become more vulnerable to pests and diseases. Due to this, our health is at a greater risk of related illnesses like diabetes and heart disease. So, we should protect our oceans and forests to ensure food for every human being.

Our environment provides us with enormous benefits which we can't repay in our entire life span. The environment includes animals, water, trees, forest and air. Trees and forests filter the air and take in harmful gases, and plants purify the water, maintain natural balance and many others. The environment keeps a regular check on its functioning as it helps regulate the vital systems essential for the ecosystem. To conclude, we can say that it is the environment that is keeping us alive. Without the blanket of the environment, we won't survive. Moreover, the environment's contribution to life cannot be repaid. Besides, what the environment has done for us, we only have damaged and degraded it.

The present book contains eleven-chapters covering all related disciplines. These chapters include Environmental science: an overview, Matter, elements, energy and organisms, Ecosystems on earth and the biosphere, Population, communities and ecosystems, Conservation of biodiversity, Effects of environmental hazards and human health, Water supply, availability and use, Food, hunger food security, Conventional farming systems and sustainable agriculture, Air pollution, global climate change & ozone layer depletion, Concepts of conventional and sustainable energy. Related terminology is given at the end for ready reference. It is written in a lucid style with the most recent advances, and with theoretical discussions being supplemented with illustrations, and tables for easy understanding of the subject. In this book most of the relevant topics are covered adequately. It offers an invaluable introduction to the living environment for all areas of study, from environmental history, agriculture and forestry, to impact assessment, climate change, ecology and conservation. The book is also supported with a comprehensive glossary and index in the end. This book will provide adequate information with a minimum of effort on the Environmental Biology to readers / students, irrespective of their discipline. The book will also be of immense use to those appearing in competitive examinations.

We are grateful to all those persons as well as various books, manuals, periodicals, magazines, journals etc. that helped in the preparation of this book. In spite of the best efforts, it is possible that some errors may have occurred into the compilation and editing of the book. Further queries, constructive suggestions and criticisms for the improvement of the book are always welcomed and shall be thankfully acknowledged.

Emily Rodriguez
Liam Roy
Isabella Martin

CONTENTS

CHAPTER-1

ENVIRONMENTAL SCIENCE: AN OVERVIEW

"Environmental Protection" by ejaugsburg is in the Public Domain, CC0

1.1 INTRODUCTION TO THE EARTH, HUMANS & ENVIRONMENT

Environmental Science?

Environmental science is the dynamic, interdisciplinary study of the interaction of living and non-living parts of the environment, with special focus on the impact of humans on the environment. The study of environmental science includes circumstances, objects, or conditions by which an organism or community is surrounded and the complex ways in which they interact.

The need for equitable, ethical, and sustainable use of Earth's resources by a global population that nears the carrying capacity of the planet requires us not only to understand how human behaviors affect the environment, but also the scientific principles that govern interactions between the living and non-living. Our future depends on our ability to understand and evaluate evidence-based arguments about the environmental consequences of human actions and technologies, and to make informed decisions based on those arguments.

From global climate change to habitat loss driven by human population growth and development, Earth is becoming a different planet—right before our eyes. The global scale and rate of environmental change are beyond anything in recorded human history. Our challenge is to acquire an improved understanding of Earth's complex environmental systems; systems characterized by interactions within and among their natural and human components that link local to global and short-term to long-term phenomena, and individual behavior to collective action. The complexity of environmental challenges demands that we all participate in finding and implementing solutions leading to long-term environmental sustainability.

A brief history of planet Earth:

1.2 WHAT IS THE PROCESS OF SCIENCE

(a) (b)

Figure 1. (a) The cyanobacteria seen through a light microscope are some of Earth's oldest life forms. These (b) stromatolites along the shores of Lake Thetis in Western Australia are ancient structures formed by the layering of cyanobacteria in shallow waters. (Credit a: modification of work by NASA; scale-bar data from Matt Russell; credit b: modification of work by Ruth Ellison)

Like other natural sciences, environmental science is a science that gathers knowledge about the natural world. The methods of science include careful observation, record keeping, logical and mathematical reasoning, experimentation, and submitting conclusions to the scrutiny of others. Science also requires considerable imagination and creativity; a well-designed experiment is commonly described as elegant or beautiful. Science has considerable practical implications and some science is dedicated to practical applications, such as the prevention of disease (figure 2). Other science proceeds largely motivated by curiosity. Whatever its goal, there is no doubt that science has transformed human existence and will continue to do so.

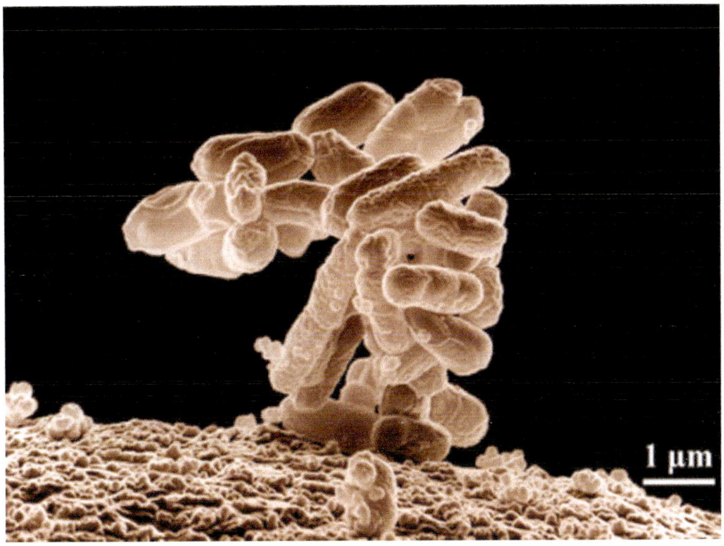

Figure 2. Biologists may choose to study Escherichia coli (E. coli), a bacterium that is a normal resident of our digestive tracts but which is also sometimes responsible for disease outbreaks. In this micrograph, the bacterium is visualized using a scanning electron microscope and digital colorization. (credit: Eric Erbe; digital colorization by Christopher Pooley, USDA-ARS)

What is the Nature of Science?

Biology is a science, but what exactly is science? What does the study of biology share with other scientific disciplines? **Science** (from the Latin *scientia,* meaning "knowledge") can be defined as a process of gaining knowledge about the natural world.

Science is a very specific way of learning about the world. The history of the past 500 years demonstrates that science is a very powerful way of gaining knowledge about the world; it is largely responsible for the technological revolutions that have taken place during this time. There are areas of knowledge, however, that the methods of science cannot be applied to. These include such things as morality, aesthetics, or spirituality. Science cannot investigate these areas because they are outside the realm of material phenomena, the phenomena of matter and energy, and cannot be observed and measured.

The **scientific method** is a method of research with defined steps that include experiments and careful observation. The steps of the scientific method will be examined in detail later, but one of the most important aspects of this method is the testing of hypotheses. A **hypothesis** is an proposed explanatory statement, for a given natural phenomenon, that can be tested. Hypotheses, or tentative explanations, are different than a **scientific theory**. A scientific theory is a widely-accepted, thoroughly tested and confirmed explanation for a set of observations or phenomena. Scientific theory is the foundation of scientific knowledge. In addition, in many scientific disciplines (less so in biology) there are **scientific laws**, often expressed in mathematical formulas, which describe how elements of nature will behave under certain specific conditions, but they do not offer explanations for why they occur.

Natural Sciences

What would you expect to see in a museum of natural sciences? Frogs? Plants? Dinosaur skeletons? Exhibits about how the brain functions? A planetarium? Gems and minerals? Or maybe all of the above? Science includes such diverse fields as astronomy, computer sciences, psychology,biology, and mathematics. However, those fields of science related to the physical world and its phenomena and processes are considered **natural sciences** and include the disciplines of physics, geology, biology, and chemistry. Environmental science is a cross-disciplinary natural science because it relies of the disciplines of chemistry, biology, and geology.

Scientific Inquiry

One thing is common to all forms of science: an ultimate goal to know. Curiosity and inquiry are the driving forces for the development of science. Scientists seek to understand the world and the way it operates. Two methods of logical thinking are used: inductive reasoning and deductive reasoning.

Inductive reasoning is a form of logical thinking that uses related observations to arrive at a general conclusion. This type of reasoning is common in descriptive science. A life scientist such as a biologist makes observations and records them. These data can be qualitative (descriptive) or quantitative (consisting of numbers), and the raw data can be supplemented with drawings, pictures, photos, or videos. From many observations, the scientist can infer conclusions (inductions) based on evidence. Inductive reasoning involves formulating generalizations inferred from careful observation and the analysis of a large amount of data. Brain studies often work this way. Many brains are observed while people are doing a task. The part of the brain that lights up, indicating activity, is then demonstrated to be the part controlling the response to that task.

Deductive reasoning or deduction is the type of logic used in hypothesis-based science. In deductive reasoning, the pattern of thinking moves in the opposite direction as compared to inductive reasoning. Deductive reasoning is a form of logical thinking that uses a general principle or law to forecast specific results. From those general principles, a scientist can extrapolate and predict the specific results that would be valid as long as the general principles are valid. For example, a prediction would be that if the climate is becoming warmer in a region, the distribution of plants and animals should change. Comparisons have been made between distributions in the past and the present, and the many changes that have been found are consistent with a warming climate. Finding the change in distribution is evidence that the climate change conclusion is a valid one.

Both types of logical thinking are related to the two main pathways of scientific study: descriptive science and hypothesis-based science. **Descriptive** (or discovery) **science** aims to observe, explore, and discover, while **hypothesis-based science** begins with a specific question or problem and a potential answer or solution that can be tested. The boundary between these two forms of study is often blurred, because most scientific endeavors combine both approaches. Observations lead to questions, questions lead to forming a hypothesis as a possible answer to those questions, and then the hypothesis is tested. Thus, descriptive science and hypothesis-based science are in continuous dialogue.

"Scientists have become the bearers of the torch of discovery in our quest for knowledge." *- Stephen Hawking and Leonard Mlodinov, in* The Grand Design *(2010), Bantam Books*

Hypothesis Testing

Biologists study the living world by posing questions about it and seeking science-based responses. This approach is common to other sciences as well and is often referred to as the scientific method. The scientific method was used even in ancient times, but it was first documented by England's Sir Francis Bacon (1561–1626) who set up inductive methods for scientific inquiry. The scientific method is not exclusively used by biologists but can be applied to almost anything as a logical problem-solving method.

The scientific process typically starts with an observation (often a problem to be solved) that leads to a question. Let's think about a simple problem that starts with an observation and apply the scientific method to solve the problem. One Monday morning, a student arrives at class and quickly discovers that the classroom is too warm. That is an observation that also describes a problem: the classroom is too warm. The student then asks a question: "Why is the classroom so warm?"

Figure 3. Sir Francis Bacon is credited with being the first to document the scientific method.

Recall that a hypothesis is a suggested explanation that can be tested. To solve a problem, several hypotheses may be proposed. For example, one hypothesis might be, "The classroom is warm because no one turned on the air conditioning." But there could be other responses to the question, and therefore other hypotheses may be proposed. A second hypothesis might be, "The classroom is warm because there is a power failure, and so the air conditioning doesn't work."

Once a hypothesis has been selected, a prediction may be made. A prediction is similar to a hypothesis but it typically has the format "If . . . then . . . ." For example, the prediction for the first hypothesis might be, "*If* the student turns on the air conditioning, *then* the classroom will no longer be too warm."

A hypothesis must be testable to ensure that it is valid. For example, a hypothesis that depends on what a bear thinks is not testable, because it can never be known what a bear thinks. It should also be **falsifiable**, meaning that it can be disproven by experimental results. An example of an unfalsifiable hypothesis is "Botticelli's *Birth of Venus* is beautiful." There is no experiment that might show this statement to be false. To test a hypothesis, a researcher will conduct one or more experiments designed to eliminate one or more of the hypotheses. This is important. A hypothesis can be disproven, or eliminated, but it can never be proven. Science does not deal in proofs like mathematics. If an experiment fails to disprove a hypothesis, then we find support for that explanation, but this is not to say that down the road a better explanation will not be found, or a more carefully designed experiment will be found to falsify the hypothesis. Each experiment will have one or more variables and one or more controls. **Experimental variables** are any part of the experiment that can vary or change during the experiment. **Controlled variables** are parts of the experiment that do not change. Lastly, experiments might have a **control group**: a group of test subjects that are as similar as possible to all other test subjects, with the exception that they don't receive the experimental treatment (those that do receive it are known as the **experimental group**). For example, in a study testing a weight-loss drug, the control group would be test subjects that don't receive the drug (but they might receive a placebo, such as sugar pill, instead). Look for these various things in the example that follows:

An experiment might be conducted to test the hypothesis that phosphate (a nutrient) promotes the growth of algae in freshwater ponds. A series of artificial ponds are filled with water and half of them are treated by adding phosphate each week, while the other half are treated by adding a non-nutritional mineral that is not used by algae. The experimental variable here is presence/absence of a nutrient (phosphate). One potential controlled variable would be the volume of water in each tank. The amount of water that algae have access to may influence the results, thus researchers want to *control* its influence on the results by making sure all test subjects get the same amount. The control group consists of the tanks that received a placebo (non-nutritional mineral) instead of the phosphate. If the ponds with phosphate show more algal growth, then we have found support for the hypothesis. If they do not, then we reject our hypothesis. Be aware that rejecting one hypothesis does not determine whether or not the other hypotheses can be accepted; it simply eliminates one hypothesis that is not valid (Figure 3). Using the scientific method, the hypotheses that are inconsistent with experimental data are rejected.

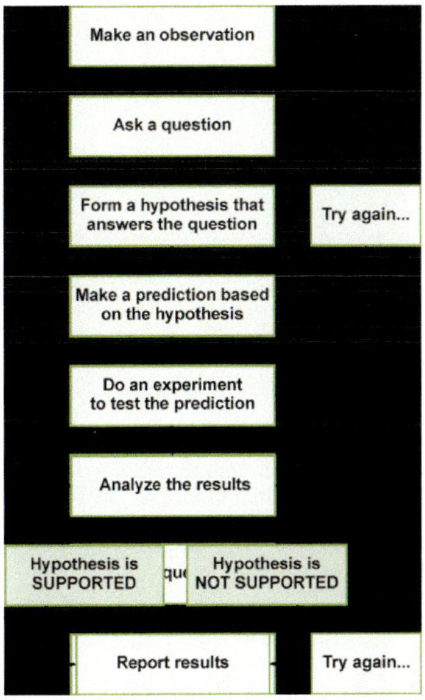

Figure 4. The scientific method is a series of defined steps that include experiments and careful observation. If a hypothesis is not supported by data, a new hypothesis can be proposed.

In the example below, the scientific method is used to solve an everyday problem. Which part in the example below is the hypothesis? Which is the prediction? Based on the results of the experiment, is the hypothesis supported? If it is not supported, propose some alternative hypotheses.

- My toaster doesn't toast my bread.
- Why doesn't my toaster work?
- There is something wrong with the electrical outlet.
- If something is wrong with the outlet, my coffeemaker also won't work when plugged into it.
- I plug my coffeemaker into the outlet.
- My coffeemaker works.

In practice, the scientific method is not as rigid and structured as it might at first appear. Sometimes an experiment leads to conclusions that favor a change in approach; often, an experiment brings entirely new scientific questions to the puzzle. Many times, science does not operate in a linear fashion; instead, scientists continually draw inferences and make generalizations, finding patterns as their research proceeds. Scientific reasoning is more complex than the scientific method alone suggests.

Basic and Applied Science

Is it valuable to pursue science for the sake of simply gaining knowledge, or does scientific knowledge only have worth if we can apply it to solving a specific problem or bettering our lives? This question focuses on the differences between two types of science: basic science and applied science.

Basic science or "pure" science seeks to expand knowledge regardless of the short-term application of that knowledge. It is not focused on developing a product or a service of immediate public or commercial value. The immediate goal of basic science is knowledge for knowledge's sake, though this does not mean that in the end it may not result in an application.

In contrast, **applied science** aims to use science to solve real-world problems, such as improving crop yield, find a cure for a particular disease, or save animals threatened by a natural disaster. In applied science, the problem is usually defined for the researcher.

Some individuals may perceive applied science as "useful" and basic science as "useless." A question these people might pose to a scientist advocating knowledge acquisition would be, "What for?" A careful look at the history of science, however, reveals that basic knowledge has resulted in many remarkable applications of great value. Many scientists think that a basic understanding of science is necessary before an application is developed; therefore, applied science relies on the results generated through basic science. Other scientists think that it is time to move on from basic science and instead to find solutions to actual problems. Both approaches are valid. It is true that there are problems that demand immediate attention; however, few solutions would be found without the help of the knowledge generated through basic science.

One example of how basic and applied science can work together to solve practical problems occurred after the discovery of DNA structure led to an understanding of the molecular mechanisms governing DNA replication. Strands of DNA, unique in every human, are found in our cells, where they provide the instructions necessary for life. During DNA replication, new copies of DNA are made, shortly before a cell divides to form new cells. Understanding the mechanisms of DNA replication (through basic science) enabled scientists to develop laboratory techniques that are now used to identify genetic diseases, pinpoint individuals who were at a crime scene, and determine paternity (all examples of applied science). Without basic science, it is unlikely that applied science would exist.

Figure 5. The Human Genome Project was a 13-year collaborative effort among researchers working in several different fields of science. The project was completed in 2003. (credit: the U.S. Department of Energy Genome Programs)

Another example of the link between basic and applied research is the Human Genome Project, a study in which each human chromosome was analyzed and mapped to determine the precise sequence of the DNA code and the exact location of each gene. (The gene is the basic unit of heredity; an individual's complete collection of genes is his or her genome.) Other organisms have also been studied as part of this project to gain a better understanding of human chromosomes. The Human Genome Project (Figure 5) relied on basic research carried out with non-human organisms and, later, with the human genome. An important end goal eventually became using the data for applied research seeking cures for genetic diseases.

Scientific Research is Transparent & Open to Critique

Whether scientific research is basic science or applied science, scientists must share their findings for other researchers to expand and build upon their discoveries. For this reason, an important aspect of a scientist's work is disseminating results and communicating with peers. Scientists can share results by presenting them at a scientific meeting or conference, but this approach can reach only the limited few who are present. Instead, most scientists present their results in peer-reviewed articles that are published in scientific journals. **Peer-reviewed articles** are scientific papers that are reviewed, usually anonymously by a scientist's colleagues, or peers. These colleagues are qualified individuals, often experts in the same research area, who judge whether or not the scientist's work is suitable for publication. The process of peer review helps to ensure that the research described in a scientific paper or grant proposal is original, significant, logical, ethical, and thorough. Scientists publish their work so other scientists can reproduce their experiments under similar or different conditions to expand on the findings. The experimental results must be consistent with the findings of other scientists.

As you review scientific information, whether in an academic setting or as part of your day-to-day life, it is important to think about the credibility of that information. You might ask yourself: has this scientific information been through the rigorous process of peer review? Are the conclusions based on available data and accepted by the larger scientific community? Scientists are inherently skeptical, especially if conclusions are not supported by evidence (and you should be too).

1.3 ENVIRONMENT AND SUSTAINABILITY

Sustainability: An Introduction

This section introduces the concept of sustainability, which refers to the sociopolitical, scientific, and cultural challenges of living within the means of the earth without significantly impairing its function.

Taking The Long View: Sustainability in Evolutionary and Ecological Perspective

Of the different forms of life that have inhabited the Earth in its three to four billion year history, 99.9% are now extinct. Against this backdrop, the human enterprise with its roughly 200,000-year history barely merits attention. As the American novelist Mark Twain once remarked, if our planet's history were to be compared to the Eiffel Tower, human history would be a mere smear on the very tip of the tower. But while modern humans (*Homo sapiens*) might be insignificant in geologic time, we are by no means insignificant in terms of our recent planetary impact. A 1986 study estimated that 40% of the product of terrestrial plant photosynthesis — the basis of the food chain for most animal and bird life — was being appropriated by humans for their use. More recent studies estimate that 25% of photosynthesis on continental shelves (coastal areas) is ultimately being used to satisfy human demand. Human appropriation of such natural resources is having a profound impact upon the wide diversity of other species that also depend on them.

Evolution normally results in the generation of new lifeforms at a rate that outstrips the extinction of other species; this results in strong biological diversity. However, scientists have evidence that, for the first observable time in evolutionary history, another species —Homo sapiens— has upset this balance to the degree that the rate of species extinction is now estimated at 10,000 times the rate of species renewal. Human beings, just one species among millions, are crowding out the other species we share the planet with. Evidence of human interference with the natural world is visible in practically every ecosystem from the presence of pollutants in the stratosphere to the artificially changed courses of the majority of river systems on the planet. It is argued that ever since we abandoned nomadic, gatherer-hunter ways of life for settled societies some 12,000 years ago, humans have continually manipulated their natural world to meet their needs. While this observation is a correct one, the rate, scale, and the nature of human-induced global change — particularly in the post-industrial period — is unprecedented in the history of life on Earth.

Primary reasons for this:

Firstly, mechanization of both industry and agriculture in the last century resulted in vastly improved labor productivity which enabled the creation of goods and services. Since then, scientific advance and technological innovation — powered by ever-increasing inputs of fossil fuels and their derivatives — have revolutionized every industry and created many new ones. The subsequent development of western consumer culture, and the satisfaction of the accompanying disposable mentality, has generated material flows of an unprecedented scale. The Wuppertal Institute estimates that humans are now responsible for moving greater amounts of matter across the planet than all natural occurrences (earthquakes, storms, etc.) put together.

Secondly, the sheer size of the human population is unprecedented. Every passing year adds another 90 million people to the planet. Even though the environmental impact varies significantly between countries (and within them), the exponential growth in human numbers, coupled with rising material expectations in a world of limited resources, has catapulted the issue of distribution to prominence. Global inequalities in resource consumption and purchasing power mark the clearest dividing line between the haves and the have-nots. It has become apparent that present patterns of production and consumption are unsustainable for a global population that is projected to reach between 12 billion by the year 2050. If ecological crises and rising social conflict are to countered, the present rates of over-consumption by a rich minority, and under-consumption by a large majority, will have to be brought into balance.

Thirdly, it is not only the rate and the scale of change but the nature of that change that is unprecedented. Human inventiveness has introduced chemicals and materials into the environment which either do not occur naturally at all, or do not occur in the ratios in which we have introduced them. These persistent chemical pollutants are believed to be causing alterations in the environment, the effects of which are only slowly manifesting themselves, and the full scale of which is beyond calculation. CFCs and PCBs are but two examples of the approximately 100,000 chemicals currently in global circulation. (Between 500 and 1,000 new chemicals are being added to this list annually.) The majority of these chemicals have not been tested for their toxicity on humans and other life forms, let alone tested for their effects in combination with other chemicals. These issues are now the subject of special UN and other intergovernmental working groups.

Concept of Sustainable Development

Our Common Future (1987), the report of the World Commission on Environment and Development, is widely credited with having popularized the concept of **sustainable development.** It defines sustainable development in the following ways...

- ...development that meets the needs of the present without compromising the ability of future generations to meet their own needs.
- ... sustainable development is not a fixed state of harmony, but rather a process of change in which the exploitation of resources, the orientation of the technological development, and institutional change are made consistent with future as well as present needs.

The concept of sustainability, however, can be traced back much farther to the oral histories of indigenous cultures. For example, the principle of inter-generational equity is captured in the Inuit saying, 'we do not inherit the Earth from our parents, we borrow it from our children'. The Native American 'Law of the Seventh Generation' is another illustration. According to this, before any major action was to be undertaken its potential consequences on the seventh generation had to be considered. For a species that at present is only 6,000 generations old and whose current political decision-makers operate on time scales of months or few years at most, the thought that other human cultures have based their decision-making systems on time scales of many decades seems wise but unfortunately inconceivable in the current political climate.

Environmental Equity

While much progress is being made to improve resource efficiency, far less progress has been made to improve resource distribution. Currently, just one-fifth of the global population is consuming three-quarters of the earth's resources (below photo). If the remaining four-fifths were to exercise their right to grow to the level of the rich minority it would result in ecological devastation. So far, global income inequalities and lack of purchasing power have prevented poorer countries from reaching the standard of living (and also resource consumption/waste emission) of the industrialized countries.

Global Consumption Inequality
24 % of the global population — mostly in the high-income countries — accounts for:

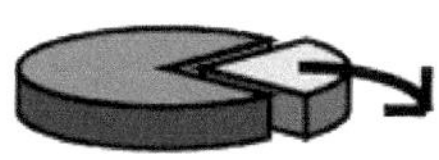

92% cars
70% CO2 emissions
86% copper and aluminium
81% paper
80% iron and steel
48% cereal crops
60% artificial fertilizer

Countries such as China, Brazil, India, and Malaysia are, however, catching up fast. In such a situation, global consumption of resources and energy needs to be drastically reduced to a point where it can be repeated by future generations. But who will do the reducing? Poorer nations want to produce and consume more. Yet so do richer countries: their economies demand ever greater consumption-based expansion. Such stalemates have prevented any meaningful progress towards equitable and sustainable resource distribution at the international level. These issue of fairness and distributional justice remain unresolved.

Basic Concepts in Environmental Science

The ecological footprint (EF), developed by Canadian ecologist and planner William Rees, is basically an accounting tool that uses land as the unit of measurement to assess per capita consumption, production, and discharge needs. It starts from the assumption that every category of energy and material consumption and waste discharge requires the productive or absorptive capacity of a finite area of land or water. If we (add up) all the land requirements for all categories of consumption and waste discharge by a defined population, the total area represents the Ecological Footprint of that population on Earth whether or not this area coincides with the population's home region.

Land is used as the unit of measurement for the simple reason that, according to Rees, "Land area not only captures planet Earth's finiteness, it can also be seen as a proxy for numerous essential life support functions from gas exchange to nutrient recycling ... land supports photosynthesis, the energy conduit for the web of life. Photosynthesis sustains all important food chains and maintains the structural integrity of ecosystems."

What does the ecological footprint tell us? Ecological footprint analysis can tell us in a vivid, ready-to-grasp manner how much of the Earth's environmental functions are needed to support human activities. It also makes visible the extent to which consumer lifestyles and behaviors are ecologically sustainable calculated that the ecological footprint of the average American is – conservatively – 5.1 hectares per capita of productive land. With roughly 7.4 billion hectares of the planet's total surface area of 51 billion hectares available for human consumption, if the current global population were to adopt American consumer lifestyles we would need two additional planets to produce the resources, absorb the wastes, and provide general life-support functions.

The **precautionary principle** is an important concept in environmental sustainability. A 1998 consensus statement characterized the precautionary principle this way: "when an activity raises threats of harm to human health or the environment, precautionary measures should be taken even if some cause and effect relationships are not fully established scientifically". For example, if a new pesticide chemical is created, the precautionary principle would dictate that we presume, for the sake of safety, that the chemical may have potential negative consequences for the environment and/or human health, even if such consequences have not been proven yet. In other words, it is best to proceed cautiously in the face of incomplete knowledge about something's potential harm.

Several Signs of Global Environmental Stress

Forests — Deforestation remains a main issue. 1 million hectares of forest were lost every year in the decade 1980-1990. The largest losses of forest area are taking place in the tropical moist deciduous forests, the zone best suited to human settlement and agriculture. Recent estimates suggest that nearly two-thirds of tropical deforestation is due to farmers clearing land for agriculture. There is increasing concern about the decline in forest quality associated with intensive use of forests and unregulated access.

Soil — As much as 10% of the earth's vegetated surface is now at least moderately degraded. Trends in soil quality and management of irrigated land raise serious questions about longer-term sustainability. It is estimated that about 20% of the world's 250 million hectares of irrigated land are already degraded to the point where crop production is seriously reduced.

Fresh Water — Some 20% of the world's population lacks access to safe water and 50% lacks access to safe sanitation. If current trends in water use persist, two-thirds of the world's population could be living in countries experiencing moderate or high water stress by 2025.

Marine fisheries — 25% of the world's marine fisheries are being fished at their maximum level of productivity and 35% are overfished (yields are declining). In order to maintain current per capita consumption of fish, global fish harvests must be increased; much of the increase might come through aquaculture which is a known source of water pollution, wetland loss and mangrove swamp destruction.

Biodiversity — Biodiversity is increasingly coming under threat from development, which destroys or degrades natural habitats, and from pollution from a variety of sources. The first comprehensive global assessment of biodiversity put the total number of species at close to 14 million and found that between 1% and 11% of the world's species may be threatened by extinction every decade. Coastal ecosystems, which host a very large proportion of marine species, are at great risk with perhaps one-third of the world's coasts at high potential risk of degradation and another 17% at moderate risk.

Atmosphere — The Intergovernmental Panel on Climate Change has established that human activities are having a discernible influence on global climate. CO_2 emissions in most industrialized countries have risen during the past few years and countries generally failed to stabilize their greenhouse gas emissions at 1990 levels by 2000 as required by the Climate Change convention.

Toxic chemicals — About 100,000 chemicals are now in commercial use and their potential impacts on human health and ecological function represent largely unknown risks. Persistent organic pollutants are now so widely distributed by air and ocean currents that they are found in the tissues of people and wildlife everywhere; they are of particular concern because of their high levels of toxicity and persistence in the environment.

Hazardous wastes — Pollution from heavy metals, especially from their use in industry and mining, is also creating serious health consequences in many parts of the world. Incidents and accidents involving uncontrolled radioactive sources continue to increase, and particular risks are posed by the legacy of contaminated areas left from military activities involving nuclear materials.

Waste — Domestic and industrial waste production continues to increase in both absolute and per capita terms, worldwide. In the developed world, per capita waste generation has increased threefold over the past 20 years; in developing countries, it is highly likely that waste generation will double during the next decade. The level of awareness regarding the health and environmental impacts of inadequate waste disposal remains rather poor; poor sanitation and waste management infrastructure is still one of the principal causes of death and disability for the urban poor.

1.4 AN OVERVIEW OF ENVIRONMENTAL ETHICS:

Frontier Ethic

The ways in which humans interact with the land and its natural resources are determined by ethical attitudes and behaviors. Early European settlers in North America rapidly consumed the natural resources of the land. After they depleted one area, they moved westward to new frontiers. Their attitude towards the land was that of a **frontier ethic**. A frontier ethic assumes that the earth has an unlimited supply of resources. If resources run out in one area, more can be found elsewhere or alternatively human ingenuity will find substitutes. This attitude sees humans as masters who manage the planet. The frontier ethic is completely anthropocentric (human-centered), for only the needs of humans are considered.

Most industrialized societies experience population and economic growth that are based upon this frontier ethic, assuming that infinite resources exist to support continued growth indefinitely. In fact, economic growth is considered a measure of how well a society is doing. The late economist Julian Simon pointed out that life on earth has never been better, and that population growth means more creative minds to solve future problems and give us an even better standard of living. However, now that the human population has passed seven billion and few frontiers are left, many are beginning to question the frontier ethic. Such people are moving toward an environmental ethic, which includes humans as part of the natural community rather than managers of it. Such an ethic places limits on human activities (e.g., uncontrolled resource use), that may adversely affect the natural community.

Some of those still subscribing to the frontier ethic suggest that outer space may be the new frontier. If we run out of resources (or space) on earth, they argue, we can simply populate other planets. This seems an unlikely solution, as even the most aggressive colonization plan would be incapable of transferring people to extraterrestrial colonies at a significant rate. Natural population growth on earth would outpace the colonization effort. A more likely scenario would be that space could provide the resources (e.g. from asteroid mining) that might help to sustain human existence on earth.

Sustainable Ethic

A **sustainable ethic** is an environmental ethic by which people treat the earth as if its resources are limited. This ethic assumes that the earth's resources are not unlimited and that humans must use and conserve resources in a manner that allows their continued use in the future. A sustainable ethic also assumes that humans are a part of the natural environment and that we suffer when the health of a natural ecosystem is impaired. A sustainable ethic includes the following tenets:

- The earth has a limited supply of resources.
- Humans must conserve resources.
- Humans share the earth's resources with other living things.
- Growth is not sustainable.
- Humans are a part of nature.
- Humans are affected by natural laws.
- Humans succeed best when they maintain the integrity of natural processes sand cooperate with nature.

For example, if a fuel shortage occurs, how can the problem be solved in a way that is consistent with a sustainable ethic? The solutions might include finding new ways to conserve oil or developing renewable energy alternatives. A sustainable ethic attitude in the face of such a problem would be that if drilling for oil damages the ecosystem, then that damage will affect the human population as well. A sustainable ethic can be either anthropocentric or biocentric (life-centered). An advocate for conserving oil resources may consider all oil resources as the property of humans. Using oil resources wisely so that future generations have access to them is an attitude consistent with an anthropocentric ethic. Using resources wisely to prevent ecological damage is in accord with a biocentric ethic.

Land Ethic

Aldo Leopold, an American wildlife natural historian and philosopher, advocated a biocentric ethic in his book, *A Sand County Almanac*. He suggested that humans had always considered land as property, just as ancient Greeks considered slaves as property. He believed that mistreatment of land (or of slaves) makes little economic or moral sense, much as today the concept of slavery is considered immoral. All humans are merely one component of an ethical framework. Leopold suggested that land be included in an ethical framework, calling this the land ethic.

"The **land ethic** simply enlarges the boundary of the community to include soils, waters, plants and animals; or collectively, the land. In short, a land ethic changes the role of *Homo sapiens* from conqueror of the land-community to plain member and citizen of it. It implies respect for his fellow members, and also respect for the community as such." (Aldo Leopold, 1949)

Leopold divided conservationists into two groups: one group that regards the soil as a commodity and the other that regards the land as biota, with a broad interpretation of its function. If we apply this idea to the field of forestry, the first group of conservationists would grow trees like cabbages, while the second group would strive to maintain a natural ecosystem. Leopold maintained that the conservation movement must be based upon more than just economic necessity. Species with no discernible economic value to humans may be an integral part of a functioning ecosystem. The land ethic respects all parts of the natural world regardless of their utility, and decisions based upon that ethic result in more stable biological communities.

"Anything is right when it tends to preserve the integrity, stability and beauty of the biotic community. It is wrong when it tends to do otherwise." (Aldo Leopold, 1949)

Hetch Hetchy Valley

In 1913, the Hetch Hetchy Valley – located in Yosemite National Park in California – was the site of a conflict between two factions, one with an anthropocentric ethic and the other, a biocentric ethic. As the last American frontiers were settled, the rate of forest destruction started to concern the public.

The conservation movement gained momentum, but quickly broke into two factions. One faction, led by Gifford Pinchot, Chief Forester under Teddy Roosevelt, advocated utilitarian conservation (i.e., conservation of resources for the good of the public). The other faction, led by John Muir, advocated preservation of forests and other wilderness for their inherent value. Both groups rejected the first tenet of frontier ethics, the assumption that resources are limitless. However, the conservationists agreed with the rest of the tenets of frontier ethics, while the preservationists agreed with the tenets of the sustainable ethic.

The Hetch Hetchy Valley was part of a protected National Park, but after the devastating fires of the 1906 San Francisco earthquake, residents of San Francisco wanted to dam the valley to provide their city with a stable supply of water. Gifford Pinchot favored the dam.

"As to my attitude regarding the proposed use of Hetch Hetchy by the city of San Francisco...I am fully persuaded that... the injury...by substituting a lake for the present swampy floor of the valley...is altogether unimportant compared with the benefits to be derived from it's use as a reservoir.

Figure 6. Yosemite valley, California, USA.

"The fundamental principle of the whole conservation policy is that of use, to take every part of the land and its resources and put it to that use in which it will serve the most people." John Muir, the founder of the Sierra Club and a great lover of wilderness, led the fight against the dam. He saw wilderness as having an intrinsic value, separate from its utilitarian value to people. He advocated preservation of wild places for their inherent beauty and for the sake of the creatures that live there. The issue aroused the American public, who were becoming increasingly alarmed at the growth of cities and the destruction of the landscape for the sake of commercial enterprises. Key senators received thousands of letters of protest.

"These temple destroyers, devotees of ravaging commercialism, seem to have a perfect contempt for Nature, and instead of lifting their eyes to the God of the Mountains, lift them to the Almighty Dollar."

Despite public protest, Congress voted to dam the valley. The preservationists lost the fight for the Hetch Hetchy Valley, but their questioning of traditional American values had some lasting effects. In 1916, Congress passed the "National

Park System Organic Act," which declared that parks were to be maintained in a manner that left them unimpaired for future generations. As we use our public lands, we continue to debate whether we should be guided by preservationism or conservationism.

Tragedy of the Commons

In his essay, The Tragedy of the Commons, Garrett Hardin (1968) looked at what happens when humans do not limit their actions by including the land as part of their ethic. The tragedy of the commons develops in the following way: Imagine a pasture open to all (the 'commons'). It is to be expected that each herdsman will try to keep as many cattle as possible on the commons. As rational beings, each herdsman seeks to maximize their gain. Adding more cattle increases their profit, and they do not suffer any immediate negative consequence because the commons are shared by all. The rational herdsman concludes that the only sensible course is to add another animal to their herd, and then another, and so forth. However, this same conclusion is reached by each and every rational herdsman sharing the commons. Therein lies the tragedy: each person is locked into a system that compels them to increase their herd, without limit, in a world that is limited. Eventually this leads to the ruination of the commons. In a society that believes in the freedom of the commons, freedom brings ruin to all because each person acts selfishly.

Hardin went on to apply the situation to modern commons: overgrazing of public lands, overuse of public forests and parks, depletion of fish populations in the ocean, use of rivers as a common dumping ground for sewage, and fouling the air with pollution.

The "Tragedy of the Commons" is applicable to what is arguably the most consequential environmental problem: global climate change. The atmosphere is a commons into which countries are dumping carbon dioxide from the burning of fossil fuels. Although we know that the generation of greenhouse gases will have damaging effects upon the entire globe, we continue to burn fossil fuels. As a country, the immediate benefit from the continued use of fossil fuels is seen as a positive component (because of economic growth). All countries, however, will share the negative long-term effects.

1.5 INDIGENOUS STRUGGLES, ENVIRONMENTAL JUSTICE, AND COMMUNITY CAPABILITIES

Environmental Justice

Environmental Justice is defined as the fair treatment and meaningful involvement of all people regardless of race, color, national origin, or income with respect to the development, implementation, and enforcement of environmental laws, regulations, and policies. It will be achieved when everyone enjoys the same degree of protection from environmental and health hazards and equal access to the decision-making process to have a healthy environment in which to live, learn, and work.

During the 1980's minority groups protested that hazardous waste sites were preferentially sited in minority neighborhoods. In 1987, Benjamin Chavis of the United Church of Christ Commission for Racism and Justice coined the term environmental racism to describe such a practice. The charges generally failed to consider whether the facility or the demography of the area came first. Most hazardous waste sites are located on property that was used as disposal sites long before modern facilities and disposal methods were available. Areas around such sites are typically depressed economically, often as a result of past disposal activities. Persons with low incomes are often constrained to live in such undesirable, but affordable, areas. The problem more likely resulted from one of insensitivity rather than racism. Indeed, the ethnic makeup of potential disposal facilities was most likely not considered when the sites were chosen.

Decisions in citing hazardous waste facilities are generally made on the basis of economics, geological suitability and the political climate. For example, a site must have a soil type and geological profile that prevents hazardous materials from moving into local aquifers. The cost of land is also an important consideration. The high cost of buying land would make it economically unfeasible to build a hazardous waste site in Beverly Hills. Some communities have seen a hazardous waste facility as a way of improving their local economy and quality of life. Emelle County, Alabama had illiteracy and infant mortality rates that were among the highest in the nation. A landfill constructed there provided jobs and revenue that ultimately helped to reduce both figures.

In an ideal world, there would be no hazardous waste facilities, but we do not live in an ideal world. Unfortunately, we live in a world plagued by rampant pollution and dumping of hazardous waste. Our industrialized society has necessarily produced wastes during the manufacture of products for our basic needs. Until technology can find a way to manage (or eliminate) hazardous waste, disposal facilities will be necessary to protect both humans and the environment. By the same token, this problem must be addressed. Industry and society must become more socially sensitive in the selection of future hazardous waste sites. All humans who help produce hazardous wastes must share the burden of dealing with those wastes, not just the poor and minorities.

Indigenous People

Since the end of the 15th century, most of the world's frontiers have been claimed and colonized by established nations. Invariably, these conquered frontiers were home to people indigenous to those regions. Some were wiped out or assimilated by the invaders, while others survived while trying to maintain their unique cultures and way of life. The United Nations officially classifies indigenous people as those "having an historical continuity with pre-invasion and pre-colonial societies," and "consider themselves distinct from other sectors of the societies now prevailing in those territories or parts of them." Furthermore, indigenous people are "determined to preserve, develop and transmit to future generations, their ancestral territories, and their ethnic identity, as the basis of their continued existence as peoples in accordance with their own cultural patterns, social institutions and legal systems." A few of the many groups of indigenous people around the world are: the many tribes of Native Americans (i.e., Navajo, Sioux) in the contiguous 48 states, the Inuit of the arctic region from Siberia to Canada, the rainforest tribes in Brazil, and the Ainu of northern Japan.

Many problems face indigenous people including the lack of human rights, exploitation of their traditional lands and themselves, and degradation of their culture. In response to the problems faced by these people, the United Nations proclaimed an "International Decade of the World's Indigenous People" beginning in 1994. The main objective of this proclamation, according to the United Nations, is "the strengthening of international cooperation for the solution of problems faced by indigenous people in such areas as human rights, the environment, development, health, culture and education." Its major goal is to protect the rights of indigenous people. Such protection would enable them to retain their cultural identity, such as their language and social customs, while participating in the political, economic and social activities of the region in which they reside.

Despite the lofty U.N. goals, the rights and feelings of indigenous people are often ignored or minimized, even by supposedly culturally sensitive developed countries. In the United States many of those in the federal government are pushing to exploit oil resources in the Arctic National Wildlife Refuge on the northern coast of Alaska. The "Gwich'in," an indigenous people who rely culturally and spiritually on the herds of caribou that live in the region, claim that drilling in the region would devastate their way of life. Thousands of years of culture would be destroyed for a few months' supply of oil. Drilling efforts have been stymied in the past, but mostly out of concern for environmental factors and not necessarily the needs of the indigenous people. Curiously, another group of indigenous people, the "Inupiat Eskimo," favor oil drilling in the Arctic National Wildlife Refuge. Because they own considerable amounts of land adjacent to the refuge, they would potentially reap economic benefits from the development of the region.

The heart of most environmental conflicts faced by governments usually involves what constitutes proper and sustainable levels of development. For many indigenous peoples, sustainable development constitutes an integrated wholeness, where no single action is separate from others. They believe that sustainable development requires the maintenance and continuity of life, from generation to generation and that humans are not isolated entities, but are part of larger communities, which include the seas, rivers, mountains, trees, fish, animals and ancestral spirits. These, along with the sun, moon and cosmos, constitute a whole. From the point of view of indigenous people, sustainable development is a process that must integrate spiritual, cultural, economic, social, political, territorial and philosophical ideals.

Figure 7. An Inupiaq woman, Nome, Alaska, c. 1907. Credit: This work is in the Public Domain, CC0

At A Glance

Science attempts to describe and understand the nature of the universe in whole or in part. Science has many fields; those fields related to the physical world and its phenomena are considered natural sciences. A hypothesis is a tentative explanation for an observation. A scientific theory is a well-tested and consistently verified explanation for a set of observations or phenomena. A scientific law is a description, often in the form of a mathematical formula, of the behavior of an aspect of nature under certain circumstances. Two types of logical reasoning are used in science. Inductive reasoning uses results to produce general scientific principles. Deductive reasoning is a form of logical thinking that predicts results by applying general principles. The common thread throughout scientific research is the use of the scientific method. Scientists present their results in peer-reviewed scientific papers published in scientific journals. Science can be basic or applied. The main goal of basic science is to expand knowledge without any expectation of short-term practical application of that knowledge. The primary goal of applied research, however, is to solve practical problems.

Sustainability refers to three simple concerns: the need to arrest environmental degradation and ecological imbalance, the need not to impoverish future generations and the need for quality of life and equity between current generations. Added up, these core concerns are an unmistakable call for transformation. Business-as-usual is no longer an option. The concept of ethics involves standards of conduct. These standards help to distinguish between behavior that is considered right and that which is considered wrong. The ways in which humans interact with the land and its natural resources are determined by ethical attitudes and behaviors. A frontier ethic assumes that the earth has an unlimited supply of resources. Environmental ethic includes humans as part of the natural community rather than managers of it. Sustainable ethic assumes that the earth's resources are not unlimited and that humans must use and conserve resources in a manner that allows their continued use in the future. Countries are categorized by a variety of methods. During the Cold War period, the United States government categorized countries according to each government's ideology and capitalistic development. Current classification models utilize economic (and sometimes other) factors in their determination. Environmental justice is achieved when everyone enjoys the same degree of protection from environmental and health hazards and equal access to the decision-making process to have a healthy environment. Many problems face indigenous people, including: lack of human rights, exploitation of their traditional lands and themselves, and degradation of their culture. Despite the lofty U.N. goals, the rights and feelings of indigenous people are often ignored or minimized, even by supposedly culturally sensitive developed countries.

CHAPTER-2

MATTER, ELEMENTS, ENERGY AND ORGANISMS

2.1 MATTER: ATOMS, MOLECULES, & COMPOUNDS

At its most fundamental level, life is made of **matter**. Matter is something that occupies space and has mass. All matter is composed of **elements**, substances that cannot be broken down or transformed chemically into other substances. Each element is made of atoms, each with a constant number of protons and unique properties. A total of 118 elements have been defined; however, only 92 occur naturally and fewer than 30 are found in living cells. The remaining 26 elements are unstable and therefore do not exist for very long or are theoretical and have yet to be detected. Each element is designated by its chemical symbol (such as H, N, O, C, and Na), and possesses unique properties. These unique properties allow elements to combine and to bond with each other in specific ways.

An atom is the smallest component of an element that retains all of the chemical properties of that element. For example, one hydrogen atom has all of the properties of the element hydrogen, such as it exists as a gas at room temperature and it bonds with oxygen to create a water molecule. Hydrogen atoms cannot be broken down into anything smaller while still retaining the properties of hydrogen. If a hydrogen atom were broken down into subatomic particles, it would no longer have the properties of hydrogen. At the most basic level, all organisms are made of a combination of elements. They contain atoms that combine together to form molecules. In multicellular organisms, such as animals, molecules can interact to form cells that combine to form tissues, which make up organs. These combinations continue until entire multicellular organisms are formed.

All matter, whether it be a rock or an organism, is made of atoms. Often, these atoms combine to form molecules. Molecule are chemicals made from two or more atoms bonded together. Some molecules are very simple, like O2, which is comprised of just two oxygen atoms. Some molecules used by organisms, such as DNA, are made of many millions of atoms. All atoms contain protons, electrons, and neutrons (Figure 1 below). The only exception is hydrogen (H), which is made of one proton and one electron. A proton is a positively charged particle that resides in the nucleus (the core of the atom) of an atom and has a mass of 1 and a charge of +1. An electron is a negatively charged particle that travels in the space around the nucleus. In other words, it resides outside of the nucleus. It has a negligible mass and has a charge of –1. Neutrons, like protons, reside in the nucleus of an atom. They have a mass of 1 and no charge. The positive (protons) and negative (electrons) charges balance each other in a neutral atom, which has a net zero charge.

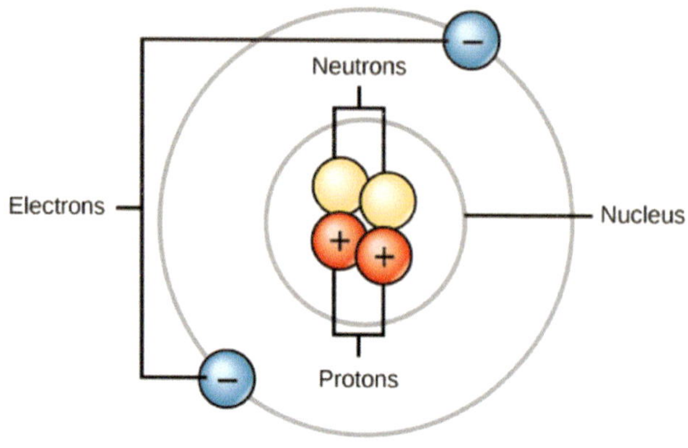

Figure 1. Atoms are comprised of protons and neutrons located within the nucleus, and electrons surrounding the nucleus. The nucleus of an atom should not be confused with the nucleus of a cell.

Each element contains a different number of protons and neutrons, giving it its own atomic number and mass number. The atomic number of an element is equal to the number of protons that element contains. The mass number is the number of protons plus the number of neutrons of that element. Therefore, it is possible to determine the number of neutrons by subtracting the atomic number from the mass number.

Isotopes are different forms of the same element that have the same number of protons, but a different number of neutrons. Some elements, such as carbon, potassium, and uranium, have naturally occurring isotopes. Carbon 12, the most common isotope of carbon, contains six protons and six neutrons. Therefore, it has a mass number of 12 (six protons and six neutrons) and an atomic number of 6 (which makes it carbon). Carbon 14 contains six protons and eight neutrons. Therefore, it has a mass number of 14 (six protons and eight neutrons) and an atomic number of 6, meaning it is still the element carbon. These two alternate forms of carbon are isotopes. Some isotopes are unstable and will lose protons, other subatomic particles, or energy to form more stable elements. These are called radioactive isotopes or radioisotopes.

Action of Evolution

Carbon dating

Carbon-14 (14C) is a naturally occurring radioisotope that is created in the atmosphere by cosmic rays. This is a continuous process, so more 14C is always being created. As a living organism develops, the relative level of 14C in its body is equal to the concentration of 14C in the atmosphere. When an organism dies, it is no longer ingesting 14C, so the ratio will decline. 14C decays to 14N by a process called beta decay; it gives off energy in this slow process. After approximately 5,730 years, only one-half of the starting concentration of 14C will have been converted to 14N. The time it takes for half of the original concentration of an isotope to decay to its more stable form is called its half-life.

Figure 2. The age of remains that contain carbon and are less than about 50,000 years old, such as this pygmy mammoth, can be determined using carbon dating. (credit: Bill Faulkner/ NPS)

Because the half-life of 14C is long, it is used to age formerly living objects, such as fossils. Using the ratio of the 14C concentration found in an object to the amount of 14C detected in the atmosphere, the amount of the isotope that has not yet decayed can be determined. Based on this amount, the age of the fossil can be calculated to about 50,000 years (Figure 2 below). Isotopes with longer half-lives, such as potassium-40, are used to calculate the ages of older fossils. Through the use of carbon dating, scientists can reconstruct the ecology and biogeography of organisms living within the past 50,000 years.

Chemical Bonds

How elements interact with one another depends on the number of electrons and how they are arranged. When an atom does not contain equal numbers of protons and electrons it is called an **ion.** Because the number of electrons does not equal the number of protons, each ion has a net **charge.** For example, if sodium loses an electron, it now has 11 protons and only 10 electrons, leaving it with an overall charge of +1. Positive ions are formed by losing electrons and are called **cations.** Negative ions are formed by gaining electrons and are called **anions.** Elemental anionic names are changed to end in -ide. As an example, when chlorine becomes an ion it is referred to as chloride.

Ionic and covalent bonds are strong bonds formed between two atoms. These bonds hold atoms together in a relatively stable state. **Ionic bonds** are formed between two oppositely charged ions (an anion and a cation). Because positive and negative charges attract, these ions are held together much like two oppositely charged magnets would stick together. **Covalent bonds** form when electrons are shared between two atoms. Each atom shares one of their electrons, which then orbits the nuclei of both atoms, holding the two atoms together. Covalent bonds are the strongest and most common form of chemical bond in organisms. Unlike most ionic bonds, covalent bonds do not dissociate in water.

Covalent bonds come in two varieties: polar and non-polar. A **non-polar covalent bond** occurs when electrons are shared equally between the two atoms. **Polar covalent bonds** form when the electrons are shared unequally. Why does this occur? Each element has a known **electronegativity**: a measure of their affinity for electrons. Some elements, such as oxygen, are very electronegative because they strongly attract electrons from other atoms. Hydrogen, meanwhile, has low electronegativity and thus weakly attracts electrons, in comparison. Polar covalent bonds form when the two atoms involved have significantly different electronegativities. In biological systems, this occurs when oxygen bonds with hydrogen and when nitrogen (also quite electronegative) bonds with hydrogen.

When oxygen and hydrogen bond, for example, the shared electrons are pulled more strongly toward oxygen and thus farther away from hydrogen's nucleus. Because the electrons move farther away from hydrogen, it becomes slightly positively charged (δ+). The oxygen becomes slightly negatively charged as the electrons become closer to it (δ-). If two molecules with polar covalent bonds approach one another, they can interact due to the attraction of opposite electrical charges. For example, the slight positive charge of hydrogen in a water molecule can be attracted to the slight negative charge of oxygen in a different water molecule (Figure 3). This interaction between two polar molecules is called a hydrogen bond. This type of bond is very common in organisms. Notably, hydrogen bonds give water the unique properties that sustain life. If it were not for hydrogen bonding, water would be a gas rather than a liquid at room temperature.

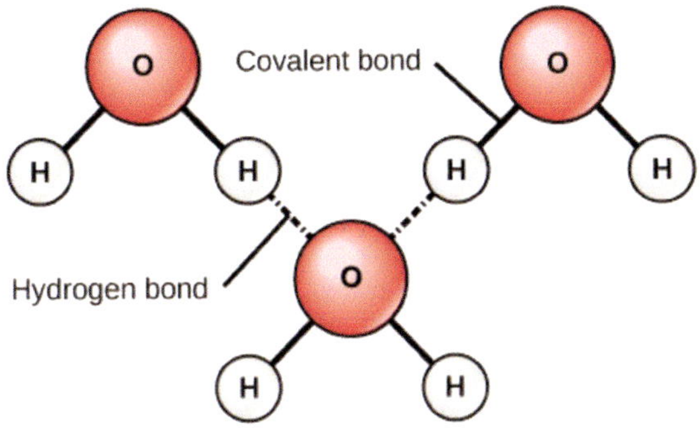

Figure 3. Hydrogen bonds form between slightly positive (δ+) and slightly negative (δ–) charges of polar covalent molecules, such as water.

2.2 MAINTENANCE OF LIFE REQUIRES WATER.

Do you ever wonder why scientists spend time looking for water on other planets? It is because water is essential to life; even minute traces of it on another planet can indicate that life could or did exist on that planet. Water is one of the more abundant molecules in living cells and the one most critical to life as we know it. Approximately 60–70 percent of your body is made up of water. Without it, life simply would not exist.

- **Water is polar:** The hydrogen and oxygen atoms within water molecules form polar covalent bonds. The shared electrons spend more time associated with the oxygen atom than they do with hydrogen atoms. There is no overall charge to a water molecule, but there is a slight positive charge on each hydrogen atom and a slight negative charge on the oxygen atom. Because of these charges, the slightly positive hydrogen atoms repel each other and form the unique shape. Each water molecule attracts other water molecules because of the positive and negative charges in the different parts of the molecule. Water also attracts other polar molecules (such as sugars) that can dissolve in water and are referred to as hydrophilic ("water-loving").

- **Water stabilizes temperature:** The hydrogen bonds in water allow it to absorb and release heat energy more slowly than many other substances. Temperature is a measure of the motion (kinetic energy) of molecules. As the motion increases, energy is higher and thus temperature is higher. Water absorbs a great deal of energy before its temperature rises. Increased energy disrupts the hydrogen bonds between water molecules. Because these bonds can be created and disrupted rapidly, water absorbs an increase in energy and temperature changes only minimally. This means that water moderates temperature changes within organisms and in their environments.

- **Water: Excellent solvent:** Because water is polar, with slight positive and negative charges, ionic compounds and polar molecules can readily dissolve in it. Water is, therefore, what is referred to as a solvent—a substance capable of dissolving another substance. The charged particles will form hydrogen bonds with a surrounding layer of water molecules.

- **Water is cohesive:** Have you ever filled up a glass of water to the very top and then slowly added a few more drops? Before it overflows, the water actually forms a dome-like shape above the rim of the glass. This water can stay above the glass because of the property of cohesion. In cohesion, water molecules are attracted to each other (because of hydrogen bonding), keeping the molecules together at the liquid-air (gas) interface, although there is no more room in the glass. Cohesion gives rise to surface tension, the capacity of a substance to withstand rupture when placed under tension or stress. When you drop a small scrap of paper onto a droplet of water, the paper floats on top of the water droplet, although the object is denser (heavier) than the water. This occurs because of the surface tension that is created by the water molecules. Cohesion and surface tension keep the water molecules intact and the item floating on the top. It is even possible to "float" a steel needle on top of a glass of water if you place it gently, without breaking the surface tension. These cohesive forces are also related to the water's property of adhesion, or the attraction between water molecules and other molecules. This is observed when water "climbs" up a straw placed in a glass of water. You will notice that the water appears to be higher on the sides of the straw than in the middle. This is because the water molecules are attracted to the straw and therefore adhere to it. Cohesive and adhesive forces are important for sustaining life. For example, because of these forces, water can flow up from the roots to the tops of plants to feed the plant.

Acids, Bases, pH, and Buffers

The pH of a solution is a measure of its acidity or alkalinity. The pH scale ranges from 0 to 14. A change of one unit on the pH scale represents a change in the concentration of hydrogen ions by a factor of 10, a change in two units represents a change in the concentration of hydrogen ions by a factor of 100. Thus, small changes in pH represent large changes in the concentrations of hydrogen ions. Pure water is neutral. It is neither acidic nor basic and has a pH of 7.0. Anything below 7.0 (ranging from 0.0 to 6.9) is acidic, and anything above 7.0 (from 7.1 to 14.0) is alkaline. The blood in your veins is slightly alkaline (pH = 7.4). The environment in your stomach is highly acidic (pH = 1 to 2). Orange juice is mildly acidic (pH = approximately 3.5), whereas baking soda is basic (pH = 9.0).

Acids are substances that provide hydrogen ions ($H+$) and lower pH, whereas **bases** provide hydroxide ions ($OH-$) and raise pH. The stronger the acid, the more readily it donates $H+$. For example, hydrochloric acid and lemon juice are very acidic and readily give up $H+$ when added to water. Conversely, bases are those substances that readily donate $OH-$. The $OH-$ ions combine with $H+$ to produce water, which raises a substance's pH. Sodium hydroxide and many household cleaners are very alkaline and give up $OH-$ rapidly when placed in water, thereby raising the pH.

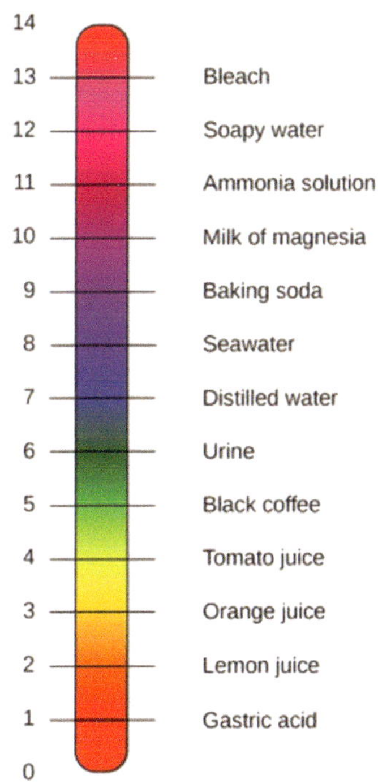

Figure 4. The pH scale measures the amount of hydrogen ions (H+) in a substance. (credit: modification of work by Edward Stevens)

How is it that we can ingest or inhale acidic or basic substances and not die? **Buffers** are the key. Buffers readily absorb excess H+ or OH–, keeping the pH of the body carefully maintained in the aforementioned narrow range. Carbon dioxide is part of a prominent buffer system in the human body; it keeps the pH within the proper range. This buffer system involves carbonic acid (H_2CO_3) and bicarbonate (HCO_3–) anion. If too much H+ enters the body, bicarbonate will combine with the H+ to create carbonic acid and limit the decrease in pH. Likewise, if too much OH– is introduced into the system, carbonic acid will combine with it to create bicarbonate and limit the increase in pH. While carbonic acid is an important product in this reaction, its presence is fleeting because the carbonic acid is released from the body as carbon dioxide gas each time we breathe. Without this buffer system, the pH in our bodies would fluctuate too much and we would fail to survive.

Biological Molecules

Besides water, the molecules necessary for life are organic.

Organic Molecules

Organic molecules are those that contain carbon covalently bonded to hydrogen. In addition, they may contain oxygen, nitrogen, phosphorus, sulfur, and additional elements. There are four major classes of organic molecules: **carbohydrates**, **lipids**, **proteins**, and **nucleic acids.** Each is an important component of the cell and performs a wide array of functions.

It is often said that life is "carbon-based." This means that carbon atoms, bonded to other carbon atoms or other elements, form the fundamental components of many of the molecules found uniquely in living things. Other elements play important roles in biological molecules, but carbon certainly qualifies as the "foundation" element for molecules in living things. It is the bonding properties of carbon atoms that are responsible for its important role.

Carbon can form four covalent bonds with other atoms or molecules. The simplest organic carbon molecule is methane (CH4), in which four hydrogen atoms bind to a carbon atom (Figure 5).

Carbohydrates

Carbohydrates include what are commonly referred to as simple sugars, like glucose, and complex carbohydrates such as starch. While many types of carbohydrates are used for energy, some are used for structure by most organisms, including plants and animals. For example, cellulose is a complex carbohydrate that adds rigidity and strength to the cell walls of plants. The suffix "-ose" denotes a carbohydrate, but note that not all carbohydrates were given that suffix when names (e.g., starch).

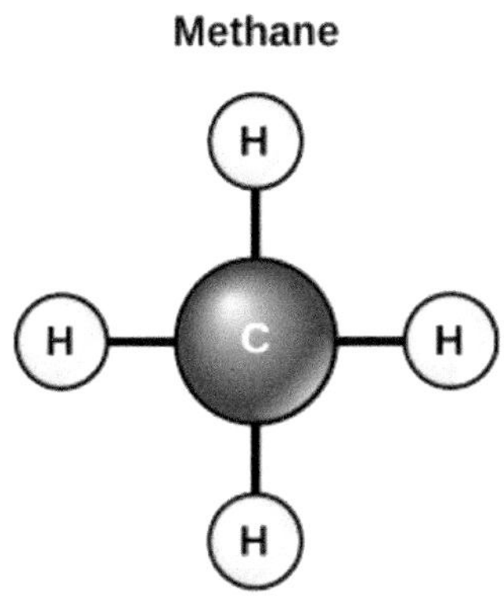

Figure 5. Carbon can form four covalent bonds to create an organic molecule. The simplest carbon molecule is methane (CH4), depicted here.

Lipids

Lipids include a diverse group of compounds that are united by a common feature. Lipids are hydrophobic ("water-fearing"), or insoluble in water, because they are non-polar molecules (molecules that contain non-polar covalent bonds) . Lipids perform many different functions in a cell. Cells store energy for long-term use in the form of lipids called fats. Lipids also provide insulation from the environment for plants and animals. For example, they help keep aquatic birds and mammals dry because of their water-repelling nature. Lipids are also the building blocks of many hormones and are an important constituent of cellular membranes. Lipids include fats, oils, waxes, phospholipids, and steroids.

Proteins

Proteins are one of the most abundant organic molecules in living systems and have the most diverse range of functions of all macromolecules. They are all polymers of amino acids. The functions of proteins are very diverse because there are 20 different chemically distinct amino acids that form long chains, and the amino acids can be in any order. Proteins can function as enzymes, hormones, contractile fibers, cytoskeleton rods, and much more.

Enzymes

Enzymes are vital to life because they act as catalyst in biochemical reactions (like digestion). Each enzyme is specific for the substrate (a reactant that binds to an enzyme) upon which it acts. Enzymes can function to break molecular bonds, to rearrange bonds, or to form new bonds.

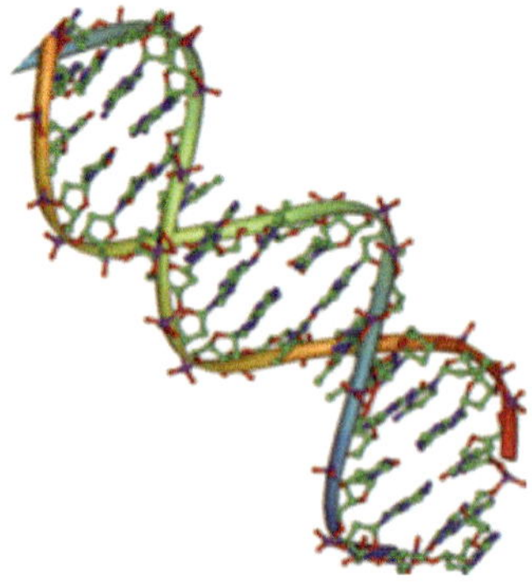

Figure 6. The double-helix model shows DNA as two parallel strands of intertwining molecules.

Nucleic Acids

Nucleic acids are very large molecules that are important to the continuity of life. They carry the genetic blueprint of a cell and thus the instructions for its functionality. The two main types of nucleic acids are deoxyribonucleic acid (**DNA**) and ribonucleic acid (RNA). DNA is the genetic material found in all organisms, ranging from single-celled bacteria to multicellular mammals. The other type of nucleic acid, **RNA,** is mostly involved in protein synthesis. The DNA molecules never leave the nucleus, but instead use an RNA intermediary to communicate with the rest of the cell. Other types of RNA are also involved in protein synthesis and its regulation. DNA and RNA are made up of small building blocks known as **nucleotides.** The nucleotides combine with each other to form a polynucleotide: DNA or RNA. Each nucleotide is made up of three components: a nitrogenous base, a pentose (five-carbon) sugar, and a phosphate. DNA has a beautiful double-helical structure (Figure 7).

Additional Resources:

Figure 7. A video that provides some wicked-awesome information about atoms.

2.2 Energy and Bioenergetics

Virtually every task performed by living organisms requires energy. Nutrients and other molecules are imported into the cell to meet these energy demands. For example, energy is required for the synthesis and breakdown of molecules, as well as the transport of molecules into and out of cells. In addition, processes such as ingesting and breaking down food, exporting wastes and toxins, and movement of the cell all require energy.

Scientists use the term bioenergetics to describe the concept of energy flow through living systems, such as cells. Cellular processes such as the building and breaking down of complex molecules occur through step-wise chemical reactions. Some of these chemical reactions are spontaneous and release energy, whereas others require energy to proceed. Together, all of the chemical reactions that take place inside cells, including those that consume or generate energy, are referred to as the cell's metabolism.

From where, and in what form, does this energy come? How do living cells obtain energy, and how do they use it? This section will discuss different forms of energy and the physical laws that govern energy transfer.

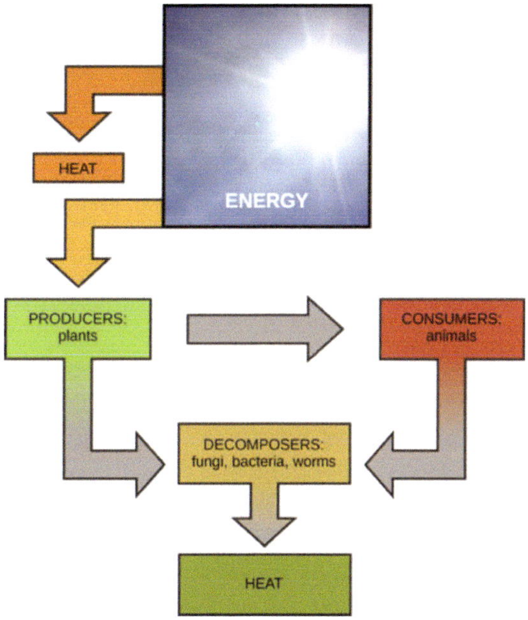

Figure 8. Ultimately, most life forms get their energy from the sun. Plants use photosynthesis to capture sunlight, and herbivores eat the plants to obtain energy. Carnivores eat the herbivores, and eventual decomposition of plant and animal material contributes to the nutrient pool.

Energy and Thermodynamics

Thermodynamics refers to the study of energy and energy transfer involving physical matter. The matter relevant to a particular case of energy transfer is called a system, and everything outside of that matter is called the surroundings. For instance, when heating a pot of water on the stove, the system includes the stove, the pot, and the water. Energy is transferred within the system (between the stove, pot, and water). There are two types of systems: open and closed. In an **open system**, energy can be exchanged with its surroundings. The stovetop system is open because heat can be lost to the air. A **closed system** cannot exchange energy with its surroundings.

Biological organisms are open systems. Energy is exchanged between them and their surroundings as they use energy from the sun to perform photosynthesis or consume energy-storing molecules and release energy to the environment by doing work and releasing heat. Like all things in the physical world, energy is subject to physical laws. The laws of thermodynamics govern the transfer of energy in and among all systems in the universe. In general, **energy** is defined as the ability to do work, or to create some kind of change. Energy exists in different forms: electrical energy, light energy, mechanical energy, and heat energy are all different types of energy. To appreciate the way energy flows into and out of biological systems, it is important to understand two of the physical laws that govern energy.

The **first law of thermodynamics** states that the total amount of energy in the universe is constant and conserved. In other words, there has always been, and always will be, exactly the same amount of energy in the universe. Energy exists in many different forms. According to the first law of thermodynamics, energy may be transferred from place to place or transformed into different forms, but it cannot be created or destroyed. The transfers and transformations of energy take place around us all the time. Light bulbs transform electrical energy into light and heat energy. Gas stoves transform chemical energy from natural gas into heat energy. Plants perform one of the most biologically useful energy transformations on earth: that of converting the energy of sunlight to chemical energy stored within organic molecules (Figure below).

The challenge for all living organisms is to obtain energy from their surroundings in forms that are usable to perform cellular work. Cells have evolved to meet this challenge. Chemical energy stored within organic molecules such as sugars and fats is transferred and transformed through a series of cellular chemical reactions into energy within molecules of ATP (adenosine triphosphate). Energy in ATP molecules is easily accessible to do work. Examples of the types of work that cells need to do include building complex molecules, transporting materials, powering the motion of cilia or flagella, and contracting muscles to create movement.

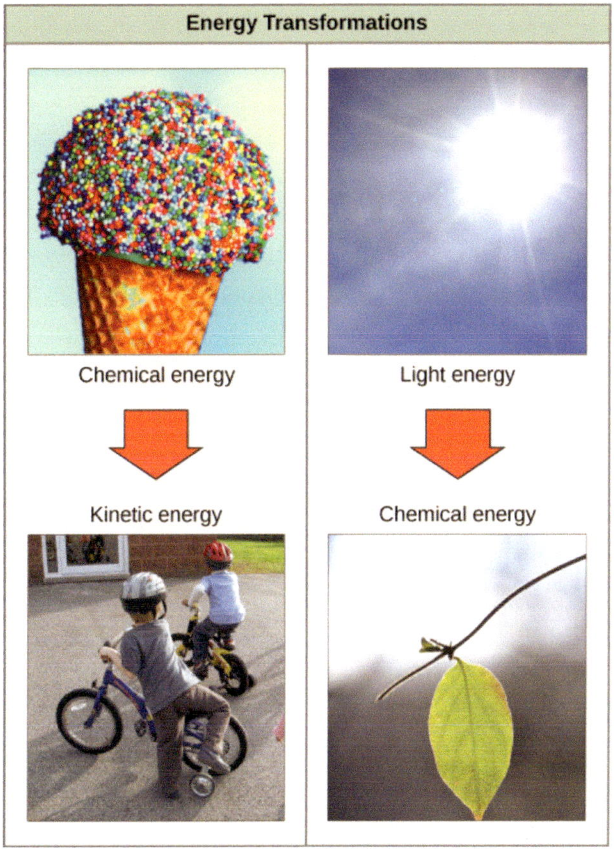

Figure 9. Shown are some examples of energy transferred and transformed from one system to another and from one form to another. The food we consume provides our cells with the energy required to carry out bodily functions, just as light energy provides plants with the means to create the chemical energy they need. (credit "ice cream": modification of work by D. Sharon Pruitt; credit "kids": modification of work by Max from Providence; credit "leaf": modification of work by Cory Zanker)

A living cell's primary tasks of obtaining, transforming, and using energy to do work may seem simple. However, the **second law of thermodynamics** explains why these tasks are harder than they appear. All energy transfers and transformations are never completely efficient. In every energy transfer, some amount of energy is lost in a form that is unusable. In most cases, this form is heat energy.

Thermodynamically, **heat energy** is defined as the energy transferred from one system to another that is not work. For example, when a light bulb is turned on, some of the energy being converted from electrical energy into light energy is lost as heat energy. Likewise, some energy is lost as heat energy during cellular metabolic reactions.

An important concept in physical systems is that of order and disorder. The more energy that is lost by a system to its surroundings, the less ordered and more random the system is. Scientists refer to the measure of randomness or disorder within a system as entropy. High entropy means high disorder and low energy. Molecules and chemical reactions have varying entropy as well. For example, entropy increases as molecules at a high concentration in one place diffuse and spread out. The second law of thermodynamics says that energy will always be lost as heat in energy transfers or transformations. Living things are highly ordered, requiring constant energy input to be maintained in a state of low entropy.

Potential Energy and Kinetic Energy

When an object is in motion, there is energy associated with that object. Think of a wrecking ball. Even a slow-moving wrecking ball can do a great deal of damage to other objects. Energy associated with objects in motion is called **kinetic energy**. A speeding bullet, a walking person, and the rapid movement of molecules in the air all have kinetic energy. Now what if that same motionless wrecking ball is lifted two stories above ground with a crane? If the suspended wrecking ball is not moving, is there energy associated with it? The answer is yes. The energy that was required to lift the wrecking ball did not disappear, but is now stored in the wrecking ball by virtue of its position and the force of gravity acting on it. This type of energy is called **potential energy** (Figure 10 below). If the ball were to fall, the potential energy would be transformed into kinetic energy until all of the potential energy was exhausted when the ball rested on the ground. Wrecking balls also swing like a pendulum; through the swing, there is a constant change of potential energy (highest at the top of the swing) to kinetic energy (highest at the bottom of the swing). Other examples of potential energy include the energy of water held behind a dam or a person about to skydive out of an airplane.

Figure 10. Still water has potential energy; moving water, such as in a waterfall or a rapidly flowing river, has kinetic energy. (credit "dam": modification of work by "Pascal"/Flickr; credit "waterfall": modification of work by Frank Gualtieri)

Potential energy is not only associated with the location of matter, but also with the structure of matter. Even a spring on the ground has potential energy if it is compressed; so does a rubber band that is pulled taut. On a molecular level, the bonds that hold the atoms of molecules together exist in a particular structure that has potential energy. The fact that energy can be released by the breakdown of certain chemical bonds implies that those bonds have potential energy. In fact, there is potential energy stored within the bonds of all the food molecules we eat, which is harnessed for use. The type of potential energy that exists within chemical bonds, and is released when those bonds are broken, is called **chemical energy**. Chemical energy is responsible for providing living cells with energy from food. The release of energy occurs when the molecular bonds within food molecules are broken.

2.3 CELL: SMALLEST UNIT OF LIFE

Levels of Biological Organization

Living things are highly organized and structured, following a hierarchy of scale from small to large (Figure 1). The atom is the smallest and most fundamental unit of matter. It consists of a nucleus surrounded by electrons. Atoms combine to form molecules, which are chemical structures consisting of at least two atoms held together by a chemical bond. In plants, animals, and many other types of organisms, molecules come together in specific ways to create structures called organelles. Organelles are small structures that exist within cells and perform specialized functions. As discussed in more detail below, all living things are made of one or more cells.

In most multicellular organisms, cells combine to make tissues, which are groups of similar cells carrying out the same function. Organs are collections of tissues grouped together based on a common function. Organs are present not only in animals but also in plants. An organ system is a higher level of organization that consists of functionally related organs. For example vertebrate animals have many organ systems, such as the circulatory system that transports blood throughout the body and to and from the lungs; it includes organs such as the heart and blood vessels. Organisms are individual living entities. For example, each tree in a forest is an organism.

All the individuals of a species living within a specific area are collectively called a population. A community is the set of different populations inhabiting a common area. For instance, all of the trees, flowers, insects, and other populations in a forest form the forest's community. The forest itself is an ecosystem. An ecosystem consists of all the living things in a particular area together with the abiotic, or non-living, parts of that environment such as nitrogen in the soil or rainwater. At the highest level of organization, the biosphere is the collection of all ecosystems, and it represents the zones of life on Earth. It includes land, water, and portions of the atmosphere.

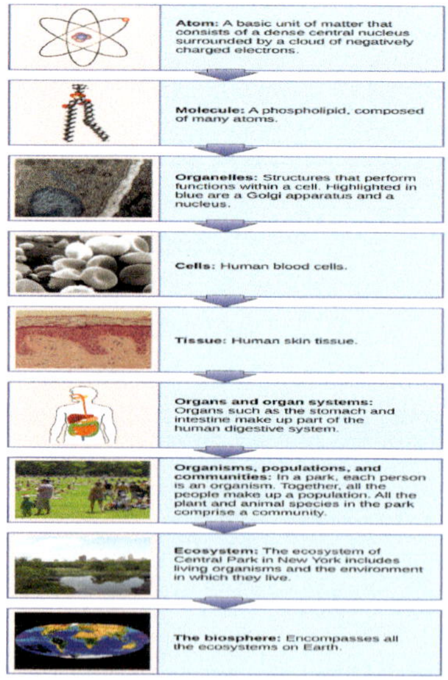

Figure 11. From an atom to the entire Earth, biology examines all aspects of life.

Fundamentals of Cell Theory

Close your eyes and picture a brick wall. What is the basic building block of that wall? It is a single brick, of course. Like a brick wall, your body is composed of basic building blocks and the building blocks of your body are cells. Your body has many kinds of cells, each specialized for a specific purpose. Just as a home is made from a variety of building materials, the human body is constructed from many cell types. For example, bone cells help to support and protect the body. Cells of the immune system fight invading bacteria. And red blood cells carry oxygen throughout the body. Each of these cell types plays a vital role during the growth, development, and day-to-day maintenance of the body. In spite of their enormous variety, however, all cells share certain fundamental characteristics.

The microscopes we use today are far more complex than those used in the 1600s by Antony van Leeuwenhoek, a Dutch shopkeeper who had great skill in crafting lenses. Despite the limitations of his now-ancient lenses, van Leeuwenhoek observed the movements of single-celled organism and sperm, which he collectively termed "animalcules." In a 1665 publication called *Micrographia*, experimental scientist Robert Hooke coined the term "cell" (from the Latin *cella*, meaning "small room") for the box-like structures he observed when viewing cork tissue through a lens. In the 1670s, van Leeuwenhoek discovered bacteria and protozoa. Later advances in lenses and microscope construction enabled other scientists to see different components inside cells.

By the late 1830s, botanist Matthias Schleiden and zoologist Theodor Schwann were studying tissues and proposed the unified **cell theory**, which states that all living things are composed of one or more cells, that the cell is the basic unit of life, and that all new cells arise from existing cells. These principles still stand today. There are many types of cells, and all are grouped into one of two broad categories: prokaryotic and eukaryotic. Animal, plant, fungal, and protist cells are classified as eukaryotic, whereas bacteria and archaea cells are classified as prokaryotic.

All cells share four common components: 1) a plasma membrane, an outer covering that separates the cell's interior from its surrounding environment; 2) cytoplasm, consisting of a jelly-like region within the cell in which other cellular components are found; 3) DNA, the genetic material of the cell; and 4) ribosomes, particles that synthesize proteins. However, prokaryotes differ from eukaryotic cells in several ways.

Prokaryotic Cells

A **prokaryotic cell** is a simple, single-celled (unicellular) organism that lacks a nucleus, or any other membrane-bound organelle. We will shortly come to see that this is significantly different in eukaryotes. Prokaryotic DNA is found in the central part of the cell: a darkened region called the nucleoid (Figure 1).

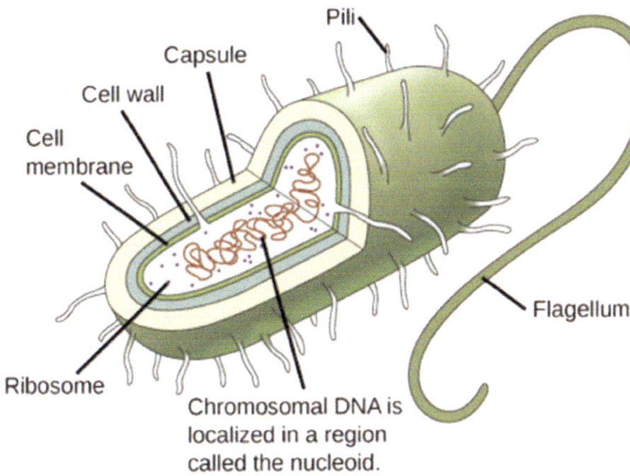

Figure 12. This figure shows the generalized structure of a prokaryotic cell.

Unlike Archaea and eukaryotes, bacteria have a cell wall made of peptidoglycan (molecules comprised of sugars and amino acids) and many have a polysaccharide capsule. The cell wall acts as an extra layer of protection, helps the cell maintain its shape, and prevents dehydration. The capsule enables the cell to attach to surfaces in its environment. Some prokaryotes have flagella, pili, or fimbriae. Flagella are used for locomotion. Pili are used to exchange genetic material during a type of reproduction called conjugation. Fimbriae are protein appendages used by bacteria to attach to other cells.

Eukaryotic Cells

A **eukaryotic cell** is a cell that has a membrane-bound nucleus and other membrane-bound compartments called **organelles.** There are many different types of organelles, each with a highly specialized function (see Figure 3). The word eukaryotic means "true kernel" or "true nucleus," alluding to the presence of the membrane-bound nucleus in these cells. The word "organelle" means "little organ," and, as already mentioned, organelles have specialized cellular functions, just as the organs of your body have specialized functions.

Cell Size

At 0.1–5.0 µm in diameter, most prokaryotic cells are significantly smaller than eukaryotic cells, which have diameters ranging from 10–100 µm (Figure 3). The small size of prokaryotes allows ions and organic molecules that enter them to quickly spread to other parts of the cell. Similarly, any wastes produced within a prokaryotic cell can quickly move out. However, larger eukaryotic cells have evolved different structural adaptations to enhance cellular transport. Indeed, the large size of these cells would not be possible without these adaptations. In general, cell size is limited because volume increases much more quickly than does cell surface area. As a cell becomes larger, it becomes more and more difficult for the cell to acquire sufficient materials to support the processes inside the cell, because the relative size of the surface area through which materials must be transported declines.

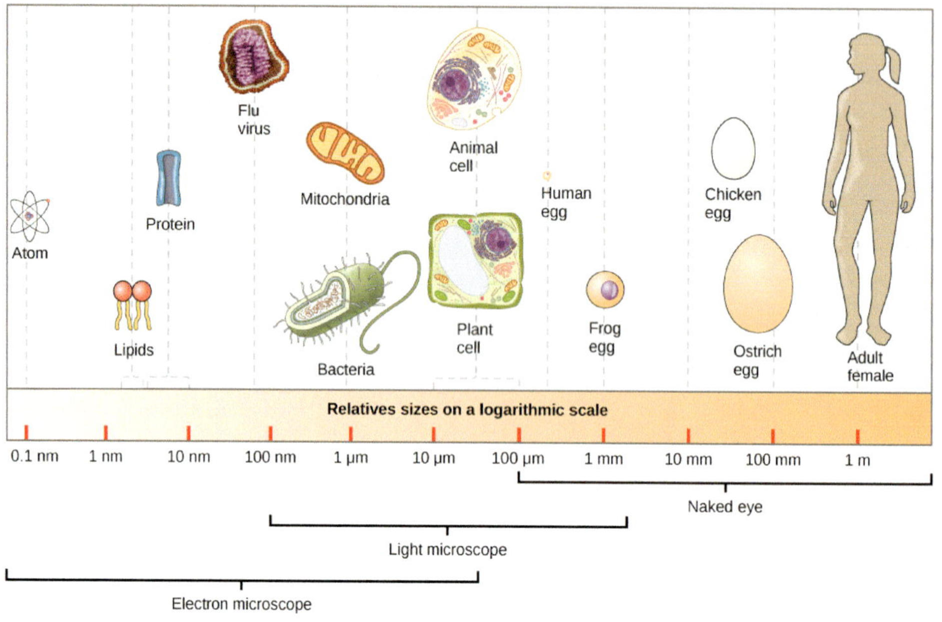

Figure 13. This figure shows the relative sizes of different kinds of cells and cellular components. An adult human is shown for comparison.

Comparison of Animal and Plant Cells

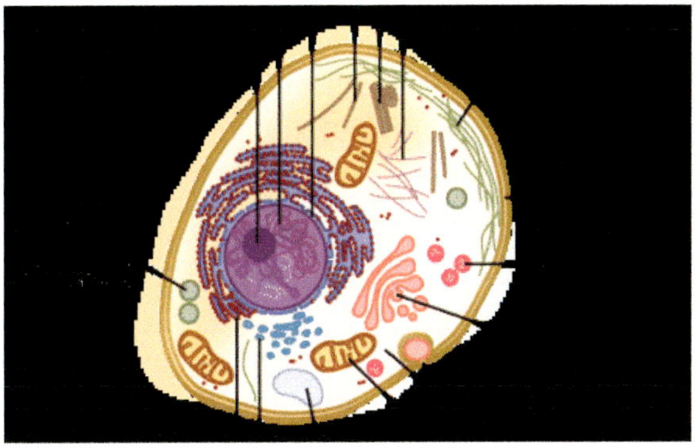

Figure 14. An example of a typical animal cell.

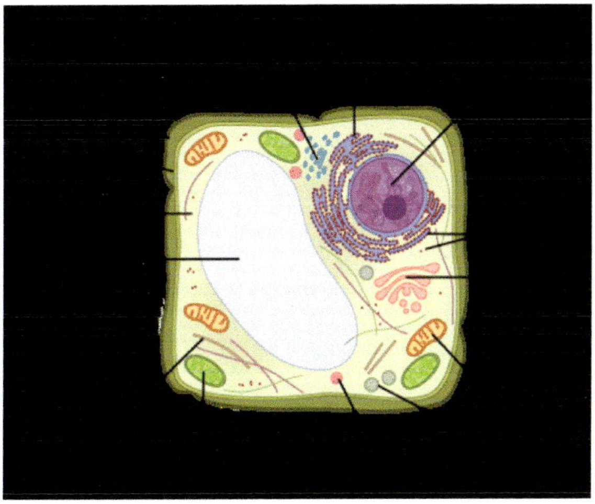

Figure 15. An example of a typical plant cell.

Despite their fundamental similarities, there are some striking differences between animal and plant cells. Animal cells have centrioles, centrosomes, and lysosomes, whereas plant cells do not. Plant cells have a rigid cell wall that is external to the plasma membrane, chloroplasts, plasmodesmata, and plastids used for storage, and a large central vacuole, whereas animal cells do not.

Chloroplasts

From an ecological perspective, **chloroplasts** are a particularly important type of organelle because they perform photosynthesis. Photosynthesis forms the foundation of food chains in most ecosystems. Chloroplasts are only found in eukaryotic cells such as plants and algae. During photosynthesis, carbon dioxide, water, and light energy are used to make glucose and molecular oxygen. One major difference between algae/plants and animals is that plants/algae are able to make their own food, like glucose, whereas animals must obtain food by consuming other organisms.

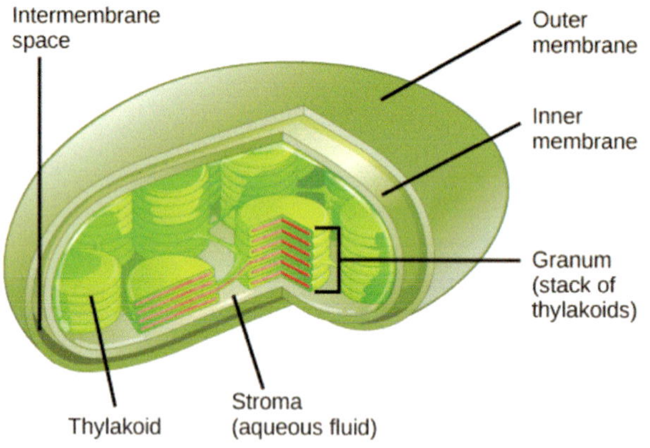

Figure 16. This simplified diagram of a chloroplast shows its structure.

Chloroplasts have outer and inner membranes, but within the space enclosed by a chloroplast's inner membrane is a set of interconnected and stacked, fluid-filled membrane sacs called thylakoids. Each stack of thylakoids is called a granum (plural = grana). The fluid enclosed by the inner membrane and surrounding the grana is called the stroma. Each structure within the chloroplast has an important function, which is enabled by its particular shape. A common theme in biology is that form and function are interrelated. For example, the membrane-rich stacks of the thylakoids provide ample surface area to embed the proteins and pigments that are vital to photosynthesis.

2.4 ENERGY, ECOSYSTEMS AND PHOTOSYNTHESIS

Cells run on the chemical energy found mainly in carbohydrate molecules, and the majority of these molecules are produced by one process: photosynthesis. Through photosynthesis, certain organisms convert solar energy (sunlight) into chemical energy, which is then used to build carbohydrate molecules. The energy stored in the bonds to hold these molecules together is released when an organism breaks down food. Cells then use this energy to perform work, such as movement. The energy that is harnessed from photosynthesis enters the ecosystems of our planet continuously and is transferred from one organism to another. Therefore, directly or indirectly, the process of photosynthesis provides most of the energy required by living things on Earth. Photosynthesis also results in the release of oxygen into the atmosphere. In short, to eat and breathe humans depend almost entirely on the organisms that carry out photosynthesis.

Solar Dependence and Food Production

(a) (b) (c)

Figure 17. (a) Plants, (b) algae, and (c) certain bacteria, called cyanobacteria, are photoautotrophs that can carry out photosynthesis. Algae can grow over enormous areas in water, at times completely covering the surface. (credit a: SteveHillebrand,U.S.FishandWildlifeService;creditb:"eutrophication&hypoxia"/ Flickr; credit c: NASA; scale-bar data from Matt Russell)

Some organisms can carry out photosynthesis, whereas others cannot. An **autotroph** is an organism that can produce its own food. The Greek roots of the word autotroph mean "self" (auto) "feeder" (troph). Plants are the best-known autotrophs, but others exist, including certain types of bacteria and algae (Figure 1). Oceanic algae contribute enormous quantities of food and oxygen to global food chains. More specifically, plants are **photoautotrophs**, a type of autotroph that uses sunlight and carbon from carbon dioxide to synthesize chemical energy in the form of carbohydrates. All organisms carrying out photosynthesis require sunlight.

Figure 18. The energy stored in carbohydrate molecules from photosynthesis passes through the food chain. The predator that eats these deer is getting energy that originated in the photosynthetic vegetation that the deer consumed. (credit: Steve VanRiper, U.S. Fish and Wildlife Service)

Heterotrophs are organisms incapable of photosynthesis that must therefore obtain energy and carbon from food by consuming other organisms. The Greek roots of the word *heterotroph* mean "other" (*hetero*) "feeder" (*troph*), meaning that their food comes from other organisms. Even if the organism being consumed is another animal, it traces its stored energy back to autotrophs and the process of photosynthesis. Humans are heterotrophs, as are all animals and fungi. Heterotrophs depend on autotrophs, either directly or indirectly. For example, a deer obtains energy by eating plants. A wolf eating a deer obtains energy that originally came from the plants eaten by that deer. Using this reasoning, all food eaten by humans can be traced back to autotrophs that carry out photosynthesis.

Photosynthesis At a Glance

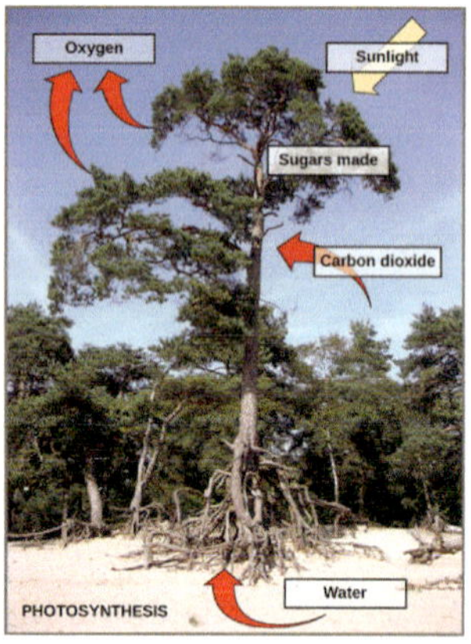

Figure 19. Photosynthesis uses solar energy, carbon dioxide, and water to release oxygen and to produce energy-storing sugar molecules.

Photosynthesis requires sunlight, carbon dioxide, and water as starting reactants (Figure 3). After the process is complete, photosynthesis releases oxygen and produces carbohydrate molecules, most commonly glucose. These sugar molecules contain the energy that living things need to survive. The complex reactions of photosynthesis can be summarized by the chemical equation shown in Figure 4.

Although the equation looks simple, the many steps that take place during photosynthesis are actually quite complex. In plants, photosynthesis takes place primarily in the chloroplasts of leaves. Chloroplasts have a double (inner and outer) membrane. Within the chloroplast is a third membrane that forms stacked, disc-shaped structures called thylakoids. Embedded in the thylakoid membrane are molecules of chlorophyll, a pigment (a molecule that absorbs light) through which the entire process of photosynthesis begins.

Photosynthesis: Two Parts

Photosynthesis takes place in two stages: the light-dependent reactions and the Calvin cycle. In the **light-dependent reactions** chlorophyll absorbs energy from sunlight and then converts it into chemical energy with the aid of water. The light-dependent reactions release **oxygen** as a byproduct from the splitting of water. In the **Calvin cycle**, the chemical energy derived from the light-dependent reactions drives both the capture of carbon in **carbon dioxide** molecules and the subsequent assembly of sugar molecules.

Photosynthesis Equation

Carbon dioxide + Water $\xrightarrow{\text{SUNLIGHT}}$ Sugar + Oxygen

Carbon dioxide	Water	Sugar	Oxygen
$6CO_2$	$6H_2O$	$C_6H_{12}O_6$	$6O_2$

Figure 20. This equation means that six molecules of carbon dioxide (CO2) combine with six molecules of water (H2O) in the presence of sunlight. This produces one molecule of glucose ($C_6H_{12}O_6$) and six molecules of oxygen (O2).

Importance of Photosynthesis

The process of photosynthesis is crucially important to the biosphere for the following reasons:

- It creates O_2, which is important for two reasons. The molecular oxygen in Earth's atmosphere was created by photosynthetic organisms; without photosynthesis there would be no O_2 to support cellular respiration needed by complex, multicellular life. Photosynthetic bacteria were likely the first organisms to perform photosynthesis, dating back 2-3 billion years ago. Thanks to their activity, and a diversity of present-day photosynthesizing organisms, Earth's atmosphere is currently about 21% O_2. Also, this O_2 is vital for the creation of the ozone layer, which protects life from harmful ultraviolet radiation emitted by the sun. Ozone (O_3) is created from the breakdown and reassembly of O_2.
- It provides energy for nearly all ecosystems. By transforming light energy into chemical energy, photosynthesis provides the energy used by organisms, whether those organisms are plants, grasshoppers, wolves, or fungi. The only exceptions are found in very rare and isolated ecosystems, such as near deep sea hydrothermal vents where organisms get energy that originally came from minerals, not the sun.
- It provides the carbon needed for organic molecules. Organisms are primarily made of two things: water and organic molecules, the latter being carbon based. Through the process of **carbon fixation,** photosynthesis takes carbon from CO_2 and converts it into sugars (which are organic).

Carbon in these sugars can be re-purposed to create the other types of organic molecules that organisms need, such as lipids, proteins, and nucleic acids. For example, the carbon used to make your DNA was once CO_2 used by photosynthetic organisms (see section 3.1 for more information on food webs).

AT A GLANCE

Matter is anything that occupies space and has mass. It is made up of atoms of different elements. Elements that occur naturally have unique qualities that allow them to combine in various ways to create compounds or molecules. Atoms, which consist of protons, neutrons, and electrons, are the smallest units of an element that retain all of the properties of that element. Electrons can be donated or shared between atoms to create bonds, including ionic, covalent, and hydrogen bonds. The pH of a solution is a measure of the concentration of hydrogen ions in the solution. Living things are carbon-based because carbon plays such a prominent role in the chemistry of living things.

A cell is the smallest unit of life. Most cells are so small that they cannot be viewed with the naked eye. The unified cell theory states that all organisms are composed of one or more cells, the cell is the basic unit of life, and new cells arise from existing cells. Each cell runs on the chemical energy found mainly in carbohydrate molecules (food), and the majority of these molecules are produced by one process: photosynthesis. Through photosynthesis, certain organisms convert solar energy (sunlight) into chemical energy, which is then used to build carbohydrate molecules. Directly or indirectly, the process of photosynthesis provides most of the energy required by living things on earth. Photosynthesis also results in the release of oxygen into the atmosphere. In short, to eat and breathe, humans depend almost entirely on the organisms that carry out photosynthesis.

CHAPTER-3

ECOSYSTEMS ON EARTH AND THE BIOSPHERE

(a) (b)

The (a) Karner blue butterfly and (b) wild lupine live in oak-pine barren habitats in North America.

This habitat is characterized by natural disturbance in the form of fire and nutrient-poor soils that are low in nitrogen—important factors in the distribution of the plants that live in this habitat. Researchers interested in ecosystem ecology study the importance of limited resources in this ecosystem and the movement of resources (such as nutrients) through the biotic and abiotic portions of the ecosystem. Researchers also examine how organisms have adapted to their ecosystem.

3.1 ENERGY FLOW AND ECOSYSTEMS

Figure 1. A (a) tidal pool ecosystem in Matinicus Island, Maine, is a small ecosystem, while the (b) Amazon rainforest in Brazil is a large ecosystem. (credit a: modification of work by Jim Kuhn; credit b: modification of work by Ivan Mlinaric)

An **ecosystem** is a community of organisms and their abiotic (non-living) environment. Ecosystems can be small, such as the tide pools found near the rocky shores of many oceans, or large, such as those found in the tropical rainforest of the Amazon in Brazil (Figure 1).

There are three broad categories of ecosystems based on their general environment: freshwater, marine, and terrestrial. Within these three categories are individual ecosystem types based on the environmental habitat and organisms present.

Freshwater ecosystems are the least common, occurring on only 1.8 percent of Earth's surface. These systems comprise lakes, rivers, streams, and springs; they are quite diverse and support a variety of animals, plants, fungi, protists and prokaryotes.

Marine ecosystems are the most common, comprising 75 percent of Earth's surface and consisting of three basic types: shallow ocean, deep ocean water, and deep ocean bottom. Shallow ocean ecosystems include extremely biodiverse coral reef ecosystems. Small photosynthetic organisms suspended in ocean waters, collectively known as **phytoplankton,** perform 40 percent of all photosynthesis on Earth. Deep ocean bottom ecosystems contain a wide variety of marine organisms. These ecosystems are so deep that light is unable to reach them.

Terrestrial ecosystems, also known for their diversity, are grouped into large categories called biomes. A **biome** is a large-scale community of organisms, primarily defined on land by the dominant plant types that exist in geographic regions of the planet with similar climatic conditions. Examples of biomes include tropical rainforests, savannas, deserts, grasslands, temperate forests, and tundras. Grouping these ecosystems into just a few biome categories obscures the great diversity of the individual ecosystems within them. For example, the saguaro cacti (*Carnegiea gigantean*) and other plant life in the Sonoran Desert, in the United States, are relatively diverse compared with the desolate rocky desert of Boa Vista, an island off the coast of Western Africa (Figure 2).

Food Chains and Food Webs

(a) (b)

Figure 2. Desert ecosystems, like all ecosystems, can vary greatly. The desert in (a) Saguaro National Park, Arizona, has abundant plant life, while the rocky desert of (b) Boa Vista island, Cape Verde, Africa, is devoid of plant life. (credit a: modification of work by Jay Galvin; credit b: modification of work by Ingo Wölbern)

A **food chain** is a linear sequence of organisms through which nutrients and energy pass as one organism eats another. The levels in the food chain are producers, primary consumers, higher-level consumers, and finally decomposers. These levels are used to describe ecosystem structure and dynamics. There is a single path through a food chain. Each organism in a food chain occupies a specific trophic level (energy level), its position in the food chain or food web.

In many ecosystems, the base, or foundation, of the food chain consists of photosynthetic organisms (plants or phytoplankton), which are called **producers.** The organisms that consume the producers are herbivores called **primary consumers. Secondary consumers** are usually carnivores that eat the primary consumers. **Tertiary consumers** are carnivores that eat other carnivores. Higher-level consumers feed on the next lower trophic levels, and so on, up to the organisms at the top of the food chain. In the Lake Ontario food chain, shown in Figure 3, the Chinook salmon is the apex consumer at the top of this food chain.

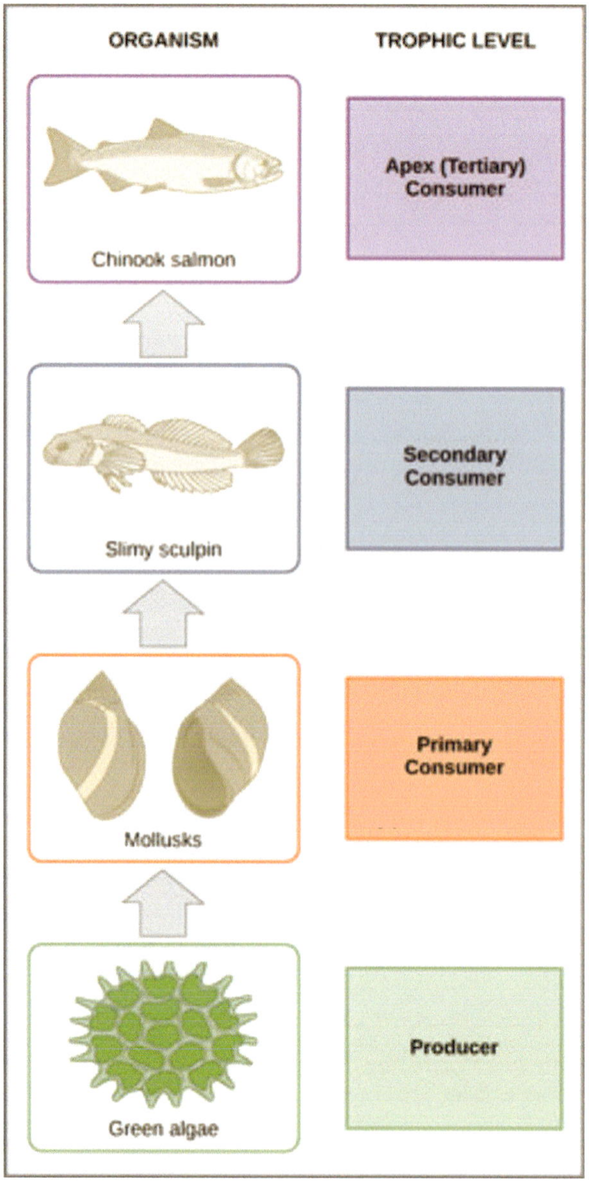

Figure 3. These are the trophic levels of a food chain in Lake Ontario at the United States–Canada border. Energy and nutrients flow from photosynthetic green algae at the base to the top of the food chain: the Chinook salmon. (credit: modification of work by National Oceanic and Atmospheric Administration/NOAA)

One major factor that limits the number of steps in a food chain is energy. Energy is lost at each trophic level and between trophic levels as heat and in the transfer to decomposers. Thus, after a limited number of trophic energy transfers, the amount of energy remaining in the food chain may not be great enough to support viable populations at higher trophic levels.

There is a one problem when using food chains to describe most ecosystems. Even when all organisms are grouped into appropriate trophic levels, some of these organisms can feed at more than one trophic level. In addition, species feed on and are eaten by more than one species. In other words, the linear model of ecosystems, the food chain, is a hypothetical and overly simplistic representation of ecosystem structure. A holistic model—which includes all the interactions between different species and their complex interconnected relationships with each other and with the environment—is a more accurate and descriptive model for ecosystems. A **food web** is a concept that accounts for the multiple trophic (feeding) interactions between each species.

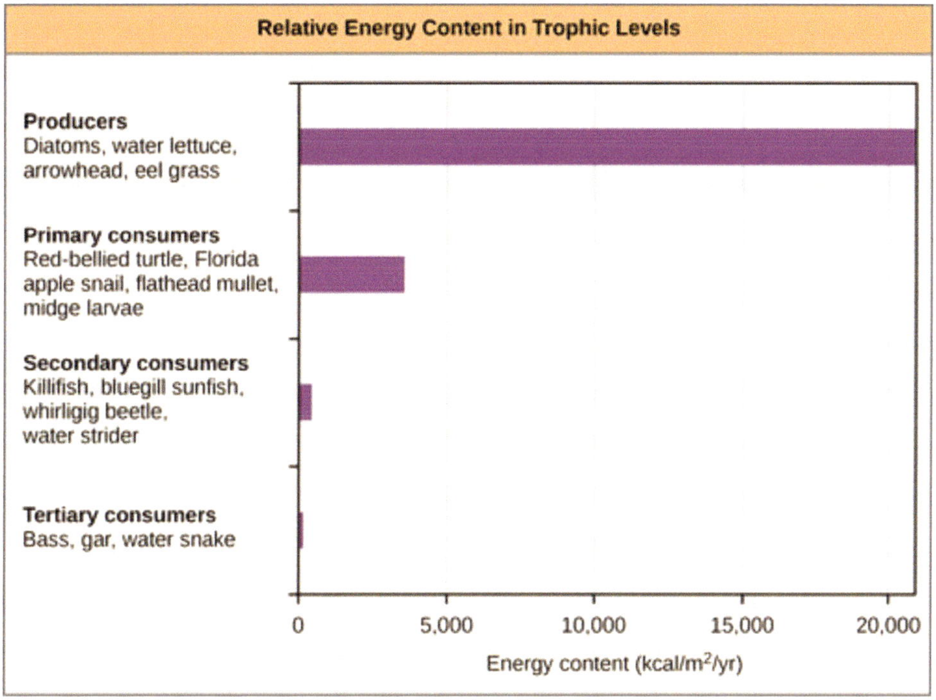

Figure 4. The relative energy in trophic levels in a Silver Springs, Florida, ecosystem is shown. Each trophic level has less energy available, and usually, but not always, supports a smaller mass of organisms at the next level.

Two general types of food webs are often shown interacting within a single ecosystem. A grazing food web has plants or other photosynthetic organisms at its base, followed by herbivores and various carnivores. A detrital food web consists of a base of organisms that feed on decaying organic matter (dead organisms), including decomposers (which break down dead and decaying organisms) and detritivores (which consume organic detritus). These organisms are usually bacteria, fungi, and invertebrate animals that recycle organic material back into the biotic part of the ecosystem as they themselves are consumed by other organisms.

Figure 5. This food web shows the interactions between organisms across trophic levels. Arrows point from an organism that is consumed to the organism that consumes it. All the producers and consumers eventually become nourishment for the decomposers (fungi, mold, earthworms, and bacteria in the soil).

Organisms Acquire Energy in a Food Web

All living things require energy in one form or another. At the cellular level, energy is used in most metabolic pathways (usually in the form of ATP), especially those responsible for building large molecules from smaller compounds. Living organisms would not be able to assemble complex organic molecules (proteins, lipids, nucleic acids, and carbohydrates) without a constant energy input.

Food-web diagrams illustrate how energy flows directionally through ecosystems. They can also indicate how efficiently organisms acquire energy, use it, and how much remains for use by other organisms of the food web. Energy is acquired by living things in two ways: autotrophs harness light or chemical energy and heterotrophs acquire energy through the consumption and digestion of other living or previously living organisms.

Photosynthetic and chemosynthetic organisms are autotrophs, which are organisms capable of synthesizing their own food (more specifically, capable of using inorganic carbon as a carbon source). Photosynthetic autotrophs (**photoautotrophs**) use sunlight as an energy source, and chemosynthetic autotrophs (**chemoautotrophs**) use inorganic molecules as an energy source. Autotrophs are critical for ecosystems because they occupy the trophic level containing producers. Without these organisms, energy would not be available to other living organisms, and life would not be possible.

Photoautotrophs, such as plants, algae, and photosynthetic bacteria, are the energy source for a majority of the world's ecosystems. Photoautotrophs harness the Sun's solar energy by converting it to chemical energy. The rate at which photosynthetic producers incorporate energy from the Sun is called **gross primary productivity**. However, not all of the energy incorporated by producers is available to the other organisms in the food web because producers must also grow and reproduce, which consumes energy. **Net primary productivity** is the energy that remains in the producers after accounting for these organisms' metabolism and heat loss. The net productivity is then available to the primary consumers at the next trophic level.

Chemoautotrophs are primarily bacteria and archaea that are found in rare ecosystems where sunlight is not available, such as those associated with dark caves or hydrothermal vents at the bottom of the ocean. Many chemoautotrophs in hydrothermal vents use hydrogen sulfide (H2S), which is released from the vents, as a source of chemical energy. This allows them to synthesize complex organic molecules, such as glucose, for their own energy and, in turn, supplies energy to the rest of the ecosystem.

One of the most important consequences of ecosystem dynamics in terms of human impact is biomagnification. **Biomagnification** is the increasing concentration of persistent, toxic substances in organisms at each successive trophic level. These are substances that are lipid soluble and are stored in the fat reserves of each organism. Many substances have been shown to biomagnify, including classical studies with the pesticide dichlorodiphenyltrichloroethane (DDT), which were described in the 1960s bestseller *Silent Spring* by Rachel Carson. DDT was a commonly used pesticide before its dangers to apex consumers, such as the bald eagle, became known. DDT and other toxins are taken in by producers and passed on to successive levels of consumers at increasingly higher rates. As bald eagles feed on contaminated fish, their DDT levels rise. It was discovered that DDT caused the eggshells of birds to become fragile, which contributed to the bald eagle being listed as an endangered species under U.S. law. The use of DDT was banned in the United States in the 1970s.

Figure 6. Swimming shrimp, a few squat lobsters, and hundreds of vent mussels are seen at a hydrothermal vent at the bottom of the ocean. As no sunlight penetrates to this depth, the ecosystem is supported by chemoautotrophic bacteria and organic material that sinks from the ocean's surface. This picture was taken in 2006 at the submerged NW Eifuku volcano off the coast of Japan by the National Oceanic and Atmospheric Administration (NOAA). The summit of this highly active volcano lies 1535 m below the surface.

Another substances that biomagnifies is polychlorinated biphenyl (PCB), which was used as coolant liquids in the United States until its use was banned in 1979. PCB was best studied in aquatic ecosystems where predatory fish species accumulated very high concentrations of the toxin that is otherwise exists at low concentrations in the environment. As illustrated in a study performed by the NOAA in the Saginaw Bay of Lake Huron of the North American Great Lakes, PCB concentrations increased from the producers of the ecosystem (phytoplankton) through the different trophic levels of fish species. The apex consumer, the walleye, has more than four times the amount of PCBs compared to phytoplankton. Also, research found that birds that eat these fish may have PCB levels that are at least ten times higher than those found in the lake fish.

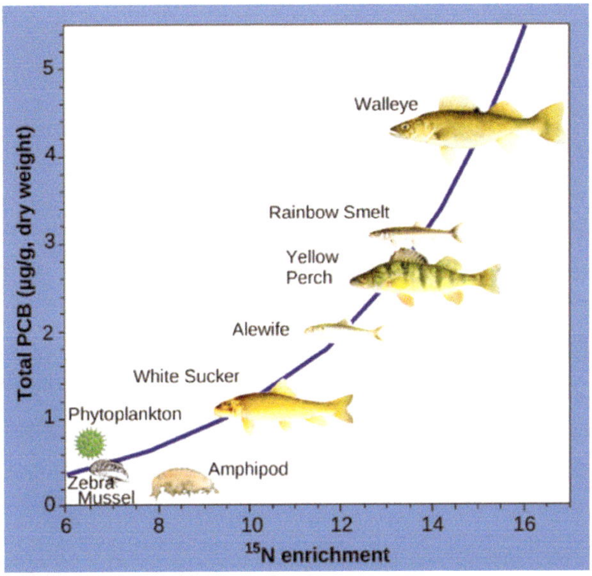

Figure 7. This chart shows the PCB concentrations found at the various trophic levels in the Saginaw Bay ecosystem of Lake Huron. Notice that the fish in the higher trophic levels accumulate more PCBs than those in lower trophic levels. (credit: Patricia Van Hoof, NOAA)

Other concerns have been raised by the biomagnification of heavy metals, such as mercury and cadmium, in certain types of seafood. The United States Environmental Protection Agency recommends that pregnant women and young children should not consume any swordfish, shark, king mackerel, or tilefish because of their high mercury content. These individuals are advised to eat fish low in mercury: salmon, shrimp, pollock, and catfish. Biomagnification is a good example of how ecosystem dynamics can affect our everyday lives, even influencing the food we eat.

3.2 BIOGEOCHEMICAL CYCLES AND ITS TYPES

Energy flows directionally through ecosystems, entering as sunlight (or inorganic molecules for chemoautotrophs) and leaving as heat during energy transformation between trophic levels. Rather than flowing through an ecosystem, the matter that makes up organisms is conserved and recycled. The six most common elements associated with organic molecules—carbon, nitrogen, hydrogen, oxygen, phosphorus, and sulfur—take a variety of chemical forms and may exist for long periods in the atmosphere, on land, in water, or beneath Earth's surface. Geologic processes, such as weathering, erosion, water drainage, and the subduction of the continental plates, all play a role in the cycling of elements on Earth. Because geology and chemistry have major roles in the study of these processes, the recycling of inorganic matter between living organisms and their nonliving environment are called **biogeochemical cycles.**

The six aforementioned elements are used by organisms in a variety of ways. Hydrogen and oxygen are found in water and organic molecules, both of which are essential to life. Carbon is found in all organic molecules, whereas nitrogen is an important component of nucleic acids and proteins. Phosphorus is used to make nucleic acids and the phospholipids that comprise biological membranes. Lastly, sulfur is critical to the three-dimensional shape of proteins.

The cycling of these elements is interconnected. For example, the movement of water is critical for the leaching of sulfur and phosphorus into rivers, lakes, and oceans. Minerals cycle through the biosphere between the biotic and abiotic components and from one organism to another.

The Water Cycle

The **hydrosphere** is the area of Earth where water movement and storage occurs: as liquid water on the surface (rivers, lakes, oceans) and beneath the surface (groundwater) or ice, (polar ice caps and glaciers), and as water vapor in the atmosphere.The human body is about 60 percent water and human cells are more than 70 percent water. Of the stores of water on Earth, 97.5 percent is salt water. Of the remaining water, more than 99 percent is groundwater or ice. Thus, less than one percent of freshwater is present in lakes and rivers. Many organisms are dependent on this small percentage, a lack of which can have negative effects on ecosystems. Humans, of course, have developed technologies to increase water availability, such as digging wells to harvest groundwater, storing rainwater, and using desalination to obtain drinkable water from the ocean. Although this pursuit of drinkable water has been ongoing throughout human history, the supply of fresh water continues to be a major issue in modern times.

The various processes that occur during the cycling of water are illustrated in Figure 8 below. The processes include the following:

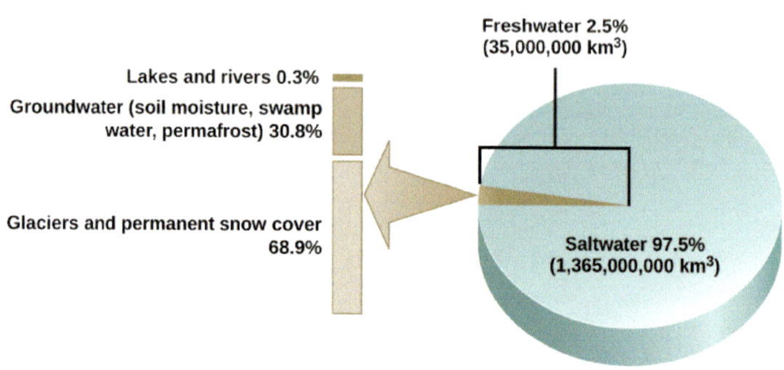

Figure 8. Only 2.5 percent of water on Earth is fresh water, and less than 1 percent of fresh water is easily accessible to living things.

- Evaporation and sublimation
- Condensation and precipitation
- Subsurface water flow
- Surface runoff and snowmelt
- Streamflow

The water cycle is driven by the Sun's energy as it warms the oceans and other surface waters. This leads to evaporation (liquid water to water vapor) of liquid surface water and sublimation (ice to water vapor) of frozen water, thus moving large amounts of water into the atmosphere as water vapor. Over time, this water vapor condenses into clouds as liquid or frozen droplets and eventually leads to precipitation (rain, snow, hail), which returns water to Earth's surface. Rain reaching Earth's surface may evaporate again, flow over the surface, or percolate into the ground. Most easily observed is surface runoff: the flow of freshwater over land either from rain or melting ice. Runoff can make its way through streams and lakes to the oceans.

In most natural terrestrial environments rain encounters vegetation before it reaches the soil surface. A significant percentage of water evaporates immediately from the surfaces of plants. What is left reaches the soil and begins to move down. Surface runoff will occur only if the soil becomes saturated with water in a heavy rainfall. Water in the soil can be taken up by plant roots. The plant will use some of this water for its own metabolism and some of that will find its way into animals that eat the plants, but much of it will be lost back to the atmosphere through a process known as transpiration: water enters the vascular system of plants through the roots and evaporates, or transpires, through the stomata (small microscope openings) of the leaves. Ecologists combine transpiration and evaporation into a single term that describes water returned to the atmosphere: evapotranspiration. Water in the soil that is not taken up by a plant and that does not evaporate is able to percolate into the subsoil and bedrock where it forms groundwater. Groundwater is a significant, subsurface reservoir of fresh water. It exists in the pores between particles in dirt, sand, and gravel or in the fissures in rocks. Groundwater can flow slowly through these pores and fissures and eventually finds its way to a stream or lake where it becomes part of the surface water again. Many streams flow not because they are replenished from rainwater directly but because they receive a constant inflow from the groundwater below. Some groundwater is found very deep in the bedrock and can persist there for millennia. Most groundwater reservoirs, or aquifers, are the source of drinking or irrigation water drawn up through wells. In many cases these aquifers are being depleted faster than they are being replenished by water percolating down from above. Rain and surface runoff are major ways in which minerals, including phosphorus and sulfur, are cycled from land to water. The environmental effects of runoff will be discussed later as these cycles are described.

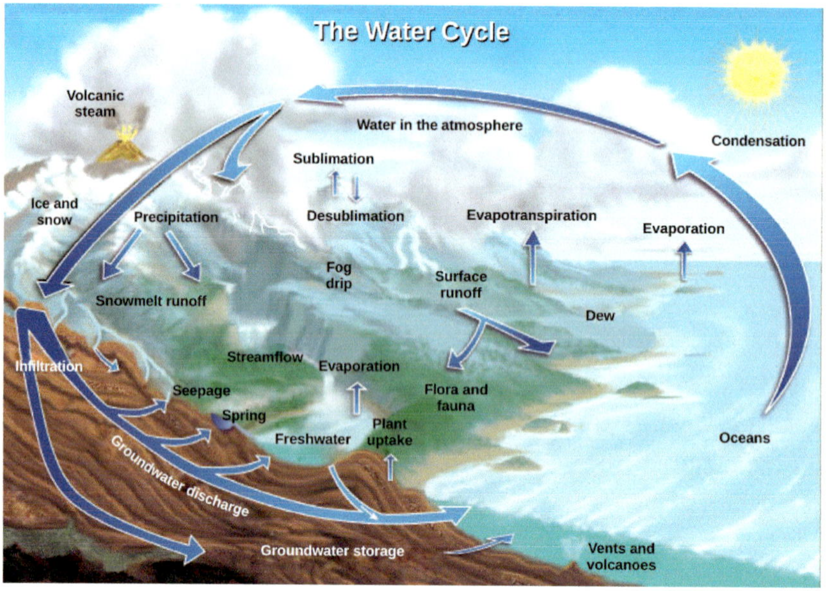

Figure 9. Water from the land and oceans enters the atmosphere by evaporation or sublimation, where it condenses into clouds and falls as rain or snow. Precipitated water may enter freshwater bodies or infiltrate the soil. The cycle is complete when surface or groundwater reenters the ocean. (credit: modification of work by John M. Evans and Howard Perlman, USGS)

The Carbon Cycle

Carbon is the second most abundant element in organisms, by mass. Carbon is present in all organic molecules (and some molecules that are not organic such as CO_2), and its role in the structure of biomolecules is of primary importance. Carbon compounds contain energy, and many of these compounds from dead plants and algae have fossilized over millions of years and are known as **fossil fuels**. Since the 1800s, the use of fossil fuels has accelerated. Since the beginning of the Industrial Revolution the demand for Earth's limited fossil fuel supplies has risen, causing the amount of carbon dioxide in our atmosphere to drastically increase. This increase in carbon dioxide is associated with climate change and is a major environmental concern worldwide.

The **carbon cycle** is most easily studied as two interconnected subcycles: one dealing with rapid carbon exchange among living organisms and the other dealing with the long-term cycling of carbon through geologic processes. The entire carbon cycle is shown in Figure below.

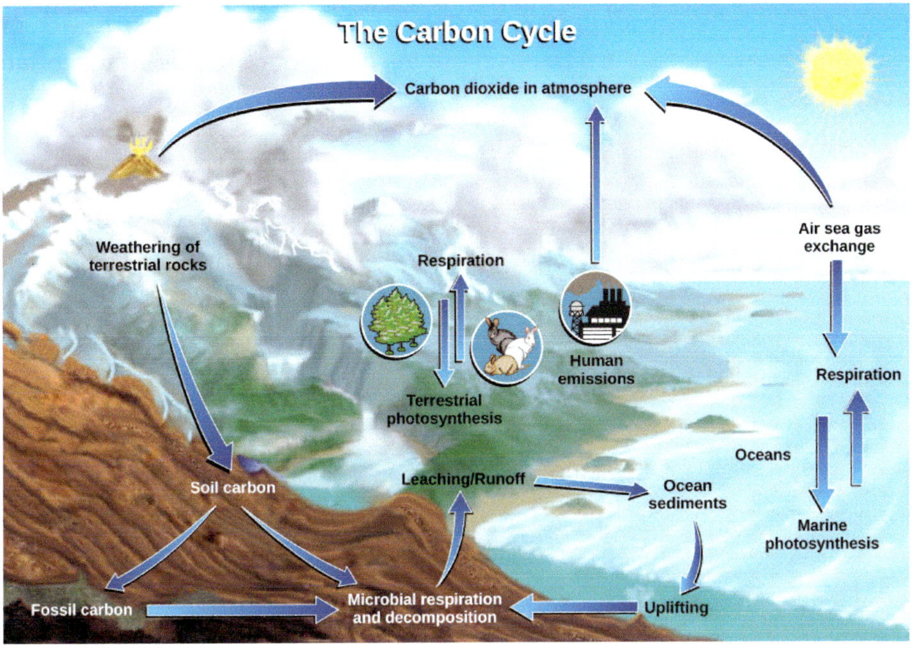

Figure 10. Carbon dioxide gas exists in the atmosphere and is dissolved in water. Photosynthesis converts carbon dioxide gas to organic carbon, and respiration cycles the organic carbon back into carbon dioxide gas. Long-term storage of organic carbon occurs when matter from living organisms is buried deep underground and becomes fossilized. Volcanic activity and, more recently, human emissions bring this stored carbon back into the carbon cycle. (credit: modification of work by John M. Evans and Howard Perlman, USGS)

The Biological Carbon Cycle

Organisms are connected in many ways, even among different ecosystems. A good example of this connection is the exchange of carbon between heterotrophs and autotrophs by way of atmospheric carbon dioxide. Carbon dioxide (CO_2) is the basic building block that autotrophs use to build high-energy compounds such as glucose. The energy harnessed from the Sun is used by these organisms to form the covalent bonds that link carbon atoms together. These chemical bonds store this energy for later use in the process of respiration. Most terrestrial autotrophs obtain their carbon dioxide directly from the atmosphere, while marine autotrophs acquire it in the dissolved form (bicarbonate, HCO_3^-).

Carbon is passed from producers to higher trophic levels through consumption. For example, when a cow (primary consumer) eats grass (producer), it obtains some of the organic molecules originally made by the plant's photosynthesis. Those organic compounds can then be passed to higher trophic levels, such as humans, when we eat the cow. At each level, however, organisms are performing **respiration**, a process in which organic molecules are broken down to release energy. As these organic molecules are broken down, carbon is removed from food molecules to form CO_2, a gas that enters the atmosphere. Thus, CO_2 is a byproduct of respiration. Recall that CO_2 is consumed by producers during photosynthesis to make organic molecules. As these molecules are broken down during respiration, the carbon once again enters the atmosphere as CO_2. Carbon exchange like this potentially connects all organisms on Earth. Think about this: the carbon in your DNA was once part of plant; millions of years ago perhaps it was part of dinosaur.

The Biogeochemical Carbon Cycle

The movement of carbon through land, water, and air is complex, and, in many cases, it occurs much more slowly than the movement between organisms. Carbon is stored for long periods in what are known as carbon reservoirs, which include the atmosphere, bodies of liquid water (mostly oceans), ocean sediment, soil, rocks (including fossil fuels), and Earth's interior.

As stated, the atmosphere is a major reservoir of carbon in the form of carbon dioxide that is essential to the process of photosynthesis. The level of carbon dioxide in the atmosphere is greatly influenced by the reservoir of carbon in the oceans. The exchange of carbon between the atmosphere and water reservoirs influences how much carbon is found in each. Carbon dioxide (CO_2) from the atmosphere dissolves in water and reacts with water molecules to form ionic compounds. Some of these ions combine with calcium ions in the seawater to form calcium carbonate ($CaCO_3$), a major component of the shells of marine organisms. These organisms eventually die and their shells form sediments on the ocean floor. Over geologic time, the calcium carbonate forms limestone, which comprises the largest carbon reservoir on Earth.

On land, carbon is stored in soil as organic carbon as a result of the decomposition of organisms or from weathering of terrestrial rock and minerals (the world's soils hold significantly more carbon than the atmosphere, for comparison). Deeper underground are fossil fuels, the anaerobically decomposed remains of plants and algae that lived millions of years ago. Fossil fuels are considered a non-renewable resource because their use far exceeds their rate of formation. A **non-renewable resource** is either regenerated very slowly or not at all. Another way for carbon to enter the atmosphere is from land (including land beneath the surface of the ocean) by the eruption of volcanoes and other geothermal systems. Carbon sediments from the ocean floor are taken deep within Earth by the process of **subduction**: the movement of one tectonic plate beneath another. Carbon is released as carbon dioxide when a volcano erupts or from volcanic hydrothermal vents.

The Nitrogen Cycle

Human activity can alter the nitrogen cycle by two primary means: the combustion of fossil fuels, which releases different nitrogen oxides, and by the use of artificial fertilizers (which contain nitrogen and phosphorus compounds) in agriculture, which are then washed into lakes, streams, and rivers by surface runoff. Atmospheric nitrogen (other than N_2) is associated with several effects on Earth's ecosystems including the production of acid rain (as nitric acid, HNO_3) and greenhouse gas effects (as nitrous oxide, N_2O), potentially causing climate change. A major effect from fertilizer runoff is saltwater and freshwater **eutrophication**, a process whereby nutrient runoff causes the overgrowth of algae, the depletion of oxygen, and death of aquatic fauna.

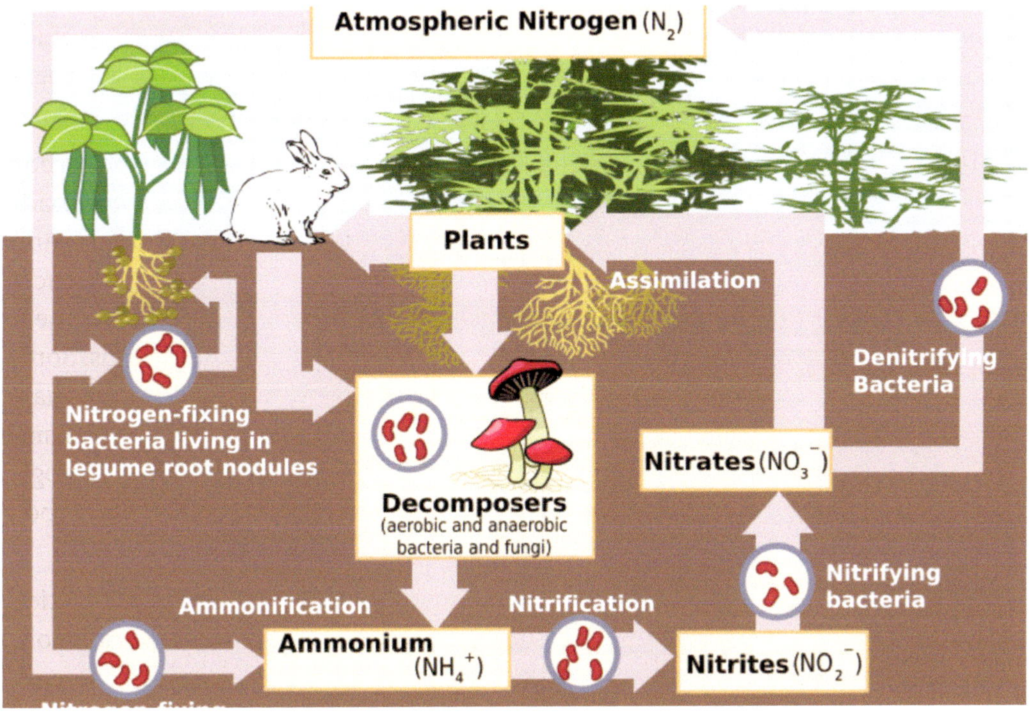

Figure 11. Nitrogen enters the living world from the atmosphere via nitrogen-fixing bacteria. This nitrogen and nitrogenous waste from animals is then processed back into gaseous nitrogen by soil bacteria, which also supply terrestrial food webs with the organic nitrogen they need. (credit: "Nitrogen cycle" by Johann Dréo & Raeky is licensed under CC BY-SA 3.0)

In marine ecosystems, nitrogen compounds created by bacteria, or through decomposition, collects in ocean floor sediments. It can then be moved to land in geologic time by uplift of Earth's crust and thereby incorporated into terrestrial rock. Although the movement of nitrogen from rock directly into living systems has been traditionally seen as insignificant compared with nitrogen fixed from the atmosphere, a recent study showed that this process may indeed be significant and should be included in any study of the global nitrogen cycle.

The Phosphorus Cycle

Phosphorus is an essential nutrient for living processes. It is a major component of nucleic acids and phospholipids, and, as calcium phosphate, it makes up the supportive components of our bones. Phosphorus is often the limiting nutrient (necessary for growth) in aquatic, particularly freshwater, ecosystems.

Phosphorus occurs in nature as the phosphate ion (PO_4^{3-}). In addition to phosphate runoff as a result of human activity, natural surface runoff occurs when it is leached from phosphate-containing rock by weathering, thus sending phosphates into rivers, lakes, and the ocean. This rock has its origins in the ocean. Phosphate-containing ocean sediments form primarily from the bodies of ocean organisms and from their excretions. However, volcanic ash, aerosols, and mineral dust may also be significant phosphate sources. This sediment then is moved to land over geologic time by the uplifting of Earth's surface. (Figure below)

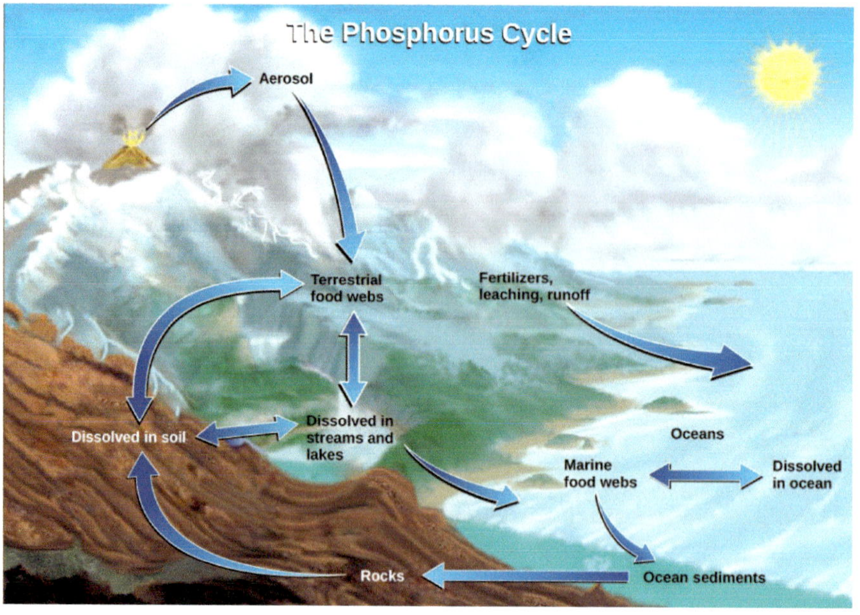

Figure 12. In nature, phosphorus exists as the phosphate ion (PO43-). Weathering of rocks and volcanic activity releases phosphate into the soil, water, and air, where it becomes available to terrestrial food webs. Phosphate enters the oceans in surface runoff, groundwater flow, and river flow. Phosphate dissolved in ocean water cycles into marine food webs. Some phosphate from the marine food webs falls to the ocean floor, where it forms sediment. (credit: modification of work by John M. Evans and Howard Perlman, USGS)

Phosphorus is also reciprocally exchanged between phosphate dissolved in the ocean and marine organisms. The movement of phosphate from the ocean to the land and through the soil is extremely slow, with the average phosphate ion having an oceanic residence time between 20,000 and 100,000 years.

Excess phosphorus and nitrogen that enter these ecosystems from fertilizer runoff and from sewage cause excessive growth of algae. The subsequent death and decay of these organisms depletes dissolved oxygen, which leads to the death of aquatic organisms such as shellfish and fish. This process is responsible for dead zones in lakes and at the mouths of many major rivers and for massive fish kills, which often occur during the summer months (see Figure 13 below).

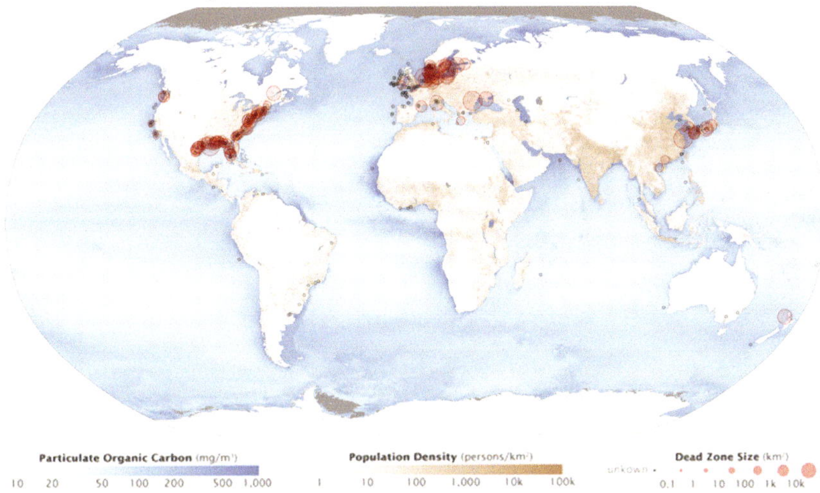

Figure 13. Dead zones occur when phosphorus and nitrogen from fertilizers cause excessive growth of microorganisms, which depletes oxygen and kills fauna. Worldwide, large dead zones are found in coastal areas of high population density. (credit: NASA Earth Observatory)

A **dead zone** is an area in lakes and oceans near the mouths of rivers where large areas are periodically depleted of their normal flora and fauna. These zones are caused by eutrophication coupled with other factors including oil spills, dumping toxic chemicals, and other human activities. The number of dead zones has increased for several years, and more than 400 of these zones were present as of 2008. One of the worst dead zones is off the coast of the United States in the Gulf of Mexico: fertilizer runoff from the Mississippi River basin created a dead zone of over 8,463 square miles. Phosphate and nitrate runoff from fertilizers also negatively affect several lake and bay ecosystems including the Chesapeake Bay in the eastern United States.

The Sulfur Cycle

Sulfur is an essential element for the molecules of living things. As part of the amino acid cysteine, it is involved in the formation of proteins. As shown in Figure 7 below, sulfur cycles between the oceans, land, and atmosphere. Atmospheric sulfur is found in the form of sulfur dioxide (SO_2), which enters the atmosphere in three ways: first, from the decomposition of organic molecules; second, from volcanic activity and geothermal vents; and, third, from the burning of fossil fuels by humans.

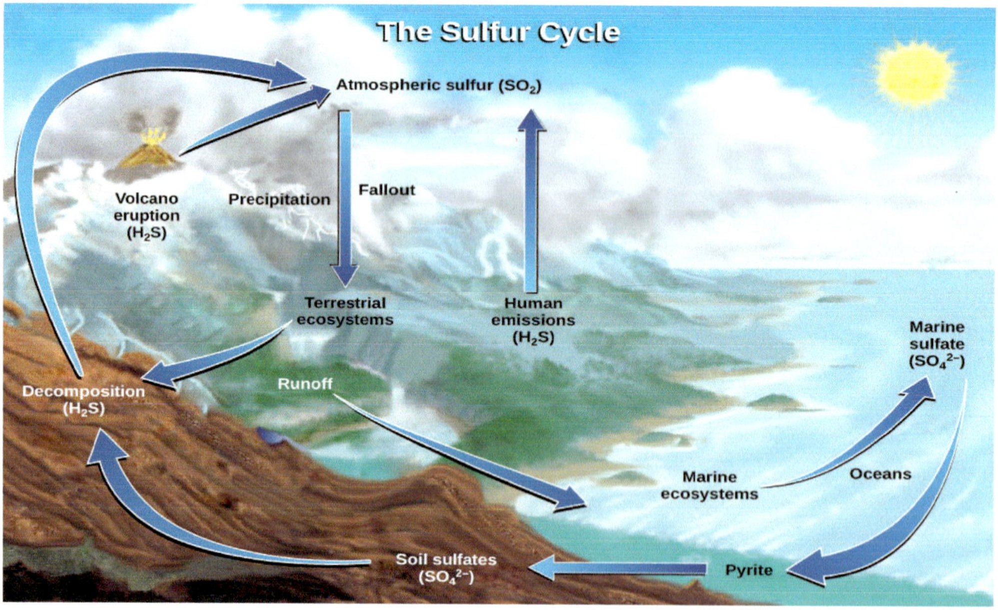

Figure 14. Sulfur dioxide from the atmosphere becomes available to terrestrial and marine ecosystems when it is dissolved in precipitation as weak sulfuric acid or when it falls directly to Earth as fallout. Weathering of rocks also makes sulfates available to terrestrial ecosystems. Decomposition of living organisms returns sulfates to the ocean, soil, and atmosphere. (credit: modification of work by John M. Evans and Howard Perlman, USGS)

On land, sulfur is deposited in four major ways: precipitation, direct fallout from the atmosphere, rock weathering, and geothermal vents. Atmospheric sulfur is found in the form of sulfur dioxide (SO_2), and as rain falls through the atmosphere, sulfur is dissolved in the form of weak sulfuric acid (H_2SO_4). Sulfur can also fall directly from the atmosphere in a process called fallout. Also, as sulfur-containing rocks weather, sulfur is released into the soil. These rocks originate from ocean sediments that are moved to land by the geologic uplifting of ocean sediments. Terrestrial ecosystems can then make use of these soil sulfates (SO_4^{2-}), which enter the food web by being taken up by plant roots. When these plants decompose and die, sulfur is released back into the atmosphere as hydrogen sulfide (H_2S) gas.

Sulfur enters the ocean in runoff from land, from atmospheric fallout, and from underwater geothermal vents. Some ecosystems rely on chemoautotrophs using sulfur as a biological energy source. This sulfur then supports marine ecosystems in the form of sulfates.

Human activities have played a major role in altering the balance of the global sulfur cycle. The burning of large quantities of fossil fuels, especially from coal, releases larger amounts of hydrogen sulfide gas into the atmosphere. As rain falls through this gas, it creates the phenomenon known as acid rain, which damages the natural environment by lowering the pH of lakes, thus killing many of the resident plants and animals. **Acid rain** is corrosive rain caused by rainwater falling to the ground through sulfur dioxide gas, turning it into weak sulfuric acid, which causes damage to aquatic ecosystems. Acid rain also affects the man-made environment through the chemical degradation of buildings. For example, many marble monuments, such as the Lincoln Memorial in Washington, DC, have suffered significant damage from acid rain over the years. These examples show the wide-ranging effects of human activities on our environment and the challenges that remain for our future.

3.3 MAJOR TERRESTRIAL BIOMES

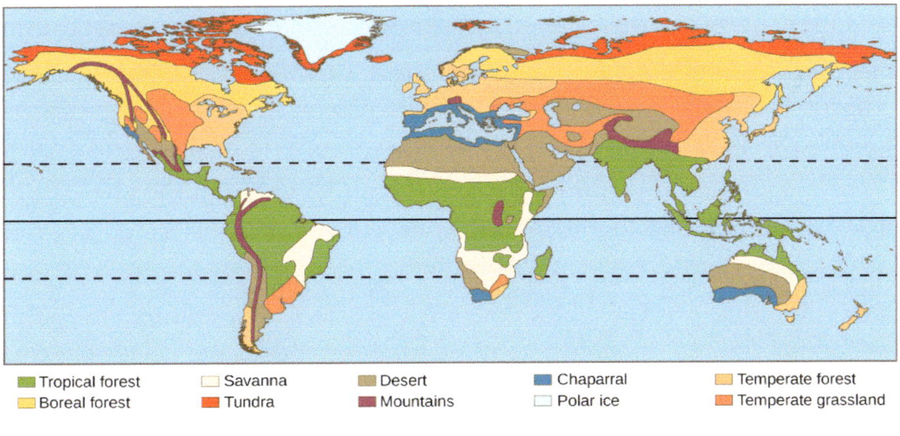

Figure 15. Each of the world's eight major biomes is distinguished by characteristic temperatures and amount of precipitation. Polar ice caps and mountains are also shown.

There are eight major terrestrial biomes: tropical rainforests, savannas, subtropical deserts, chaparral, temperate grasslands, temperate forests, boreal forests, and Arctic tundra. Biomes are large-scale environments that are distinguished by characteristic temperature ranges and amounts of precipitation. These two variables affect the types of vegetation and animal life that can exist in those areas. Because each biome is defined by climate, the same biome can occur in geographically distinct areas with similar climates.

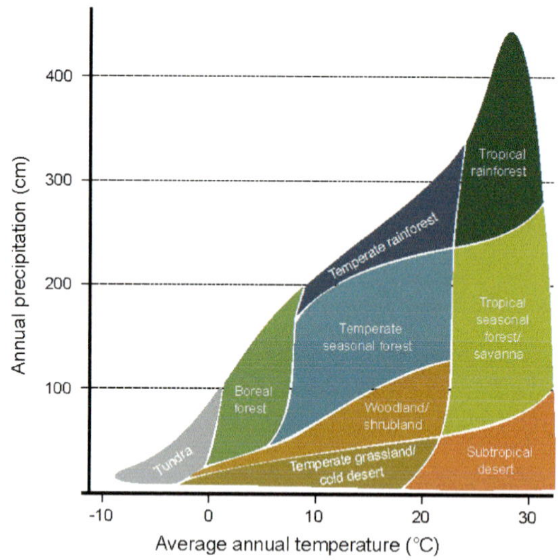

Figure 16. Precipitation and temperature are the two most important climatic variables that determine the type of biome in a particular location. Credit: "Climate influence on terrestrial biome" by Navarras is in the Public Domain, CC0

Tropical rainforests are found in equatorial regions (Figure 16) are the most biodiverse terrestrial biome. This biodiversity is under extraordinary threat primarily through logging and deforestation for agriculture. Tropical rainforests have also been described as nature's pharmacy because of the potential for new drugs that is largely hidden in the chemicals produced by the huge diversity of plants, animals, and other organisms. The vegetation is characterized by plants with spreading roots and broad leaves that fall off throughout the year, unlike the trees of deciduous forests that lose their leaves in one season.

The temperature and sunlight profiles of tropical rainforests are stable in comparison to other terrestrial biomes, with average temperatures ranging from 20°C to 34°C (68°F to 93°F). Month-to-month temperatures are relatively constant in tropical rainforests, in contrast to forests farther from the equator. This lack of temperature seasonality leads to year-round plant growth rather than just seasonal growth. In contrast to other ecosystems, a consistent daily amount of sunlight (11–12 hours per day year-round) provides more solar radiation and therefore more opportunity for primary productivity.

The annual rainfall in tropical rainforests ranges from 125 to 660 cm (50–200 in) with considerable seasonal variation. Tropical rainforests have wet months in which there can be more than 30 cm (11–12 in) of precipitation, as well as dry months in which there are fewer than 10 cm (3.5 in) of rainfall. However, the driest month of a tropical rainforest can still exceed the annual rainfall of some other biomes, such as deserts. Tropical rainforests have high net primary productivity because the annual temperatures and precipitation values support rapid plant growth. However, the high amounts of rainfall leaches nutrients from the soils of these forests.

Figure 17. Species diversity is very high in tropical wet forests, such as these forests of Madre de Dios, Peru, near the Amazon River. (credit: Roosevelt Garcia)

Tropical rainforests are characterized by vertical layering of vegetation and the formation of distinct habitats for animals within each layer. On the forest floor is a sparse layer of plants and decaying plant matter. Above that is an understory of short, shrubby foliage. A layer of trees rises above this understory and is topped by a closed upper canopy—the uppermost overhead layer of branches and leaves. Some additional trees emerge through this closed upper canopy. These layers provide diverse and complex habitats for the variety of plants, animals, and other organisms. Many species of animals use the variety of plants and the complex structure of the tropical wet forests for food and shelter. Some organisms live several meters above ground, rarely descending to the forest floor.

Figure 18. A MinuteEarth video about how trees create rainfall, and vice versa.

Savannas are grasslands with scattered trees and are found in Africa, South America, and northern Australia. Savannas are hot, tropical areas with temperatures averaging from 24°C –29°C (75°F –84°F) and an annual rainfall of 51–127 cm (20–50 in). Savannas have an extensive dry season and consequent fires. As a result, there are relatively few trees scattered in the grasses and forbs (herbaceous flowering plants) that dominate the savanna. Because fire is an important source of disturbance in this biome, plants have evolved well-developed root systems that allow them to quickly re-sprout after a fire.

Figure 19. Although savannas are dominated by grasses, small woodlands, such as this one in Mount Archer National Park in Queensland, Australia, may dot the landscape. (credit: "Ethel Aardvark"/Wikimedia Commons)

Subtropical deserts exist between 15° and 30° north and south latitude and are centered on the Tropic of Cancer and the Tropic of Capricorn (Figure 6 below). Deserts are frequently located on the downwind or lee side of mountain ranges, which create a rain shadow after prevailing winds drop their water content on the mountains. This is typical of the North American deserts, such as the Mohave and Sonoran deserts. Deserts in other regions, such as the Sahara Desert in northern Africa or the Namib Desert in southwestern Africa are dry because of the high-pressure, dry air descending at those latitudes. Subtropical deserts are very dry; evaporation typically exceeds precipitation. Subtropical hot deserts can have daytime soil surface temperatures above 60°C (140°F) and nighttime temperatures approaching 0°C (32°F). Subtropical deserts are characterized by low annual precipitation of fewer than 30 cm (12 in) with little monthly variation and lack of predictability in rainfall. Some years may receive tiny amounts of rainfall, while others receive more. In some cases, the annual rainfall can be as low as 2 cm (0.8 in) in subtropical deserts located in central Australia ("the Outback") and northern Africa.

Figure 20. A MinuteEarth video about the global climate patterns which lead to subtropical deserts.

The low species diversity of this biome is closely related to its low and unpredictable precipitation. Despite the relatively low diversity, desert species exhibit fascinating adaptations to the harshness of their environment. Very dry deserts lack perennial vegetation that lives from one year to the next; instead, many plants are annuals that grow quickly and reproduce when rainfall does occur, then they die. Perennial plants in deserts are characterized by adaptations that conserve water: deep roots, reduced foliage, and water-storing stems (Figure 6 below). Seed plants in the desert produce seeds that can lie dormant for extended periods between rains. Most animal life in subtropical deserts has adapted to a nocturnal life, spending the hot daytime hours beneath the ground. The Namib Desert is the oldest on the planet, and has probably been dry for more than 55 million years. It supports a number of endemic species (species found only there) because of this great age. For example, the unusual gymnosperm *Welwitschia mirabilis* is the only extant species of an entire order of plants. There are also five species of reptiles considered endemic to the Namib.

In addition to subtropical deserts there are **cold deserts** that experience freezing temperatures during the winter and any precipitation is in the form of snowfall. The largest of these deserts are the Gobi Desert in northern China and southern Mongolia, the Taklimakan Desert in western China, the Turkestan Desert, and the Great Basin Desert of the United States.

The chaparral is also called scrub forest and is found in California, along the Mediterranean Sea, and along the southern coast of Australia (Figure 7 below). The annual rainfall in this biome ranges from 65 cm to 75 cm (25.6–29.5 in) and the majority of the rain falls in the winter. Summers are very dry and many chaparral plants are dormant during the summertime. The chaparral vegetation is dominated by shrubs and is adapted to periodic fires, with some plants producing seeds that germinate only after a hot fire. The ashes left behind after a fire are rich in nutrients like nitrogen and fertilize the soil, promoting plant regrowth. Fire is a natural part of the maintenance of this biome.

Figure 21. Many desert plants have tiny leaves or no leaves at all to reduce water loss. The leaves of ocotillo, shown here in the Chihuahuan Desert in Big Bend National Park, Texas, appear only after rainfall and then are shed. (credit "bare ocotillo": "Leaflet"/Wikimedia Commons)

Temperate grasslands are found throughout central North America, where they are also known as prairies, and in Eurasia, where they are known as steppes (Figure 8 below). Temperate grasslands have pronounced annual fluctuations in temperature with hot summers and cold winters. The annual temperature variation produces specific growing seasons for plants. Plant growth is possible when temperatures are warm enough to sustain plant growth, which occurs in the spring, summer, and fall.

Figure 22. The chaparral is dominated by shrubs. (credit: Miguel Vieira)

Annual precipitation ranges from 25.4 cm to 88.9 cm (10–35 in). Temperate grasslands have few trees except for those found growing along rivers or streams. The dominant vegetation tends to consist of grasses. The treeless condition is maintained by low precipitation, frequent fires, and grazing. The vegetation is very dense and the soils are fertile because the subsurface of the soil is packed with the roots and rhizomes (underground stems) of these grasses. The roots and rhizomes act to anchor plants into the ground and replenish the organic material (humus) in the soil when they die and decay.

Figure 23. The American bison (*Bison bison*), more commonly called the buffalo, is a grazing mammal that once populated American prairies in huge numbers. (credit: Jack Dykinga, USDA ARS)

Fires, which are a natural disturbance in temperate grasslands, can be ignited by lightning strikes. It also appears that the lightning-caused fire regime in North American grasslands was enhanced by intentional burning by humans. When fire is suppressed in temperate grasslands, the vegetation eventually converts to scrub and dense forests. Often, the restoration or management of temperate grasslands requires the use of controlled burns to suppress the growth of trees and maintain the grasses.

Temperate forests are the most common biome in eastern North America, Western Europe, Eastern Asia, Chile, and New Zealand (Figure 9 below). This biome is found throughout mid-latitude regions. Temperatures range between –30°C and 30°C (–22°F to 86°F) and drop to below freezing on an annual basis. These temperatures mean that temperate forests have defined growing seasons during the spring, summer, and early fall. Precipitation is relatively constant throughout the year and ranges between 75 cm and 150 cm (29.5–59 in).

Deciduous trees are the dominant plant in this biome with fewer evergreen conifers. Deciduous trees lose their leaves each fall and remain leafless in the winter. Thus, little photosynthesis occurs during the dormant winter period.

Each spring, new leaves appear as temperature increases. Because of the dormant period, the net primary productivity of temperate forests is less than that of tropical rainforests. In addition, temperate forests show far less diversity of tree species than tropical rainforest biomes.

The trees of the temperate forests leaf out and shade much of the ground. However, more sunlight reaches the ground in this biome than in tropical rainforests because trees in temperate forests do not grow as tall as the trees in tropical rainforests. The soils of the temperate forests are rich in inorganic and organic nutrients compared to tropical rainforests. This is because of the thick layer of leaf litter on forest floors and reduced leaching of nutrients by rainfall. As this leaf litter decays, nutrients are returned to the soil. The leaf litter also protects soil from erosion, insulates the ground, and provides habitats for invertebrates and their predators.

The **boreal forest,** also known as **taiga** or **coniferous forest**, is found roughly between 50°and 60° north latitude across most of Canada, Alaska, Russia, and northern Europe (Figure 10 below). Boreal forests are also found above a certain elevation (and below high elevations where trees cannot grow) in mountain ranges throughout the Northern Hemisphere. This biome has cold, dry winters and short, cool, wet summers. The annual precipitation is from 40 cm to 100 cm (39–15.7 in) and usually takes the form of snow; relatively little evaporation occurs because of the cool temperatures.

Figure 24. Deciduous trees are the dominant plant in the temperate forest. (credit: Oliver Herold)

The long and cold winters in the boreal forest have led to the predominance of cold-tolerant cone-bearing plants. These are evergreen coniferous trees like pines, spruce, and fir, which retain their needle-shaped leaves year-round. Evergreen trees can photosynthesize earlier in the spring than deciduous trees because less energy from the Sun is required to warm a needle-like leaf than a broad leaf. Evergreen trees grow faster than deciduous trees in the boreal forest. In addition, soils in boreal forest regions tend to be acidic with little available nitrogen. Leaves are a nitrogen-rich structure and deciduous trees must produce a new set of these nitrogen-rich structures each year. Therefore, coniferous trees that retain nitrogen-rich needles in a nitrogen limiting environment may have had a competitive advantage over the broad-leafed deciduous trees.

Figure 25. The boreal forest (taiga) has low lying plants and conifer trees. (credit: L.B. Brubaker, NOAA)

The net primary productivity of boreal forests is lower than that of temperate forests and tropical wet forests. The aboveground biomass of boreal forests is high because these slow-growing tree species are long-lived and accumulate standing biomass over time. Species diversity is less than that seen in temperate forests and tropical rainforests. Boreal forests lack the layered forest structure seen in tropical rainforests or, to a lesser degree, temperate forests. The structure of a boreal forest is often only a tree layer and a ground layer. When conifer needles are dropped, they decompose more slowly than broad leaves; therefore, fewer nutrients are returned to the soil to fuel plant growth.

The Arctic **tundra** lies north of the subarctic boreal forests and is located throughout the Arctic regions of the Northern Hemisphere. Tundra also exists at elevations above the tree line on mountains. The average winter temperature is –°34C (–°29.2F) and the average summer temperature is °3C–°12C (°37F –°52F). Plants in the Arctic tundra have a short growing season of approximately 60–50 days. However, during this time, there are almost 24 hours of daylight and plant growth is rapid. The annual precipitation of the Arctic tundra is low (25–15 cm or 10–6 in) with little annual variation in precipitation. And, as in the boreal forests, there is little evaporation because of the cold temperatures.

Figure 26. Low-growing plants such lichen and grasses are common in tundra. Credit: Nunavut tundra by Flickr: My Nunavut is licensed under CC BY 2.0

Plants in the Arctic tundra are generally low to the ground and include low shrubs, grasses, lichens, and small flowering plants (Figure 11 below). There is little species diversity, low net primary productivity, and low above-ground biomass. The soils of the Arctic tundra may remain in a perennially frozen state referred to as permafrost. The permafrost makes it impossible for roots to penetrate far into the soil and slows the decay of organic matter, which inhibits the release of nutrients from organic matter. The melting of the permafrost in the brief summer provides water for a burst of productivity while temperatures and long days permit it. During the growing season, the ground of the Arctic tundra can be completely covered with plants or lichens.

3.4 OVERVIEW OF AQUATIC BIOMES

Aquatic Biomes are influenced by abiotic factors.

Like terrestrial biomes, **aquatic biomes** are influenced by a series of abiotic factors. The aquatic medium—water— has different physical and chemical properties than air. Even if the water in a pond or other body of water is perfectly clear (there are no suspended particles), water still absorbs light. As one descends into a deep body of water, there will eventually be a depth which the sunlight cannot reach. While there are some abiotic and biotic factors in a terrestrial ecosystem that might obscure light (like fog, dust, or insect swarms), usually these are not permanent features of the environment. The importance of light in aquatic biomes is central to the communities of organisms found in both freshwater and marine ecosystems. In freshwater systems, stratification due to differences in density is perhaps the most critical abiotic factor and is related to the energy aspects of light. The thermal properties of water (rates of heating and cooling) are significant to the function of marine systems and have major impacts on global climate and weather patterns. Marine systems are also influenced by large-scale physical water movements, such as currents; these are less important in most freshwater lakes.

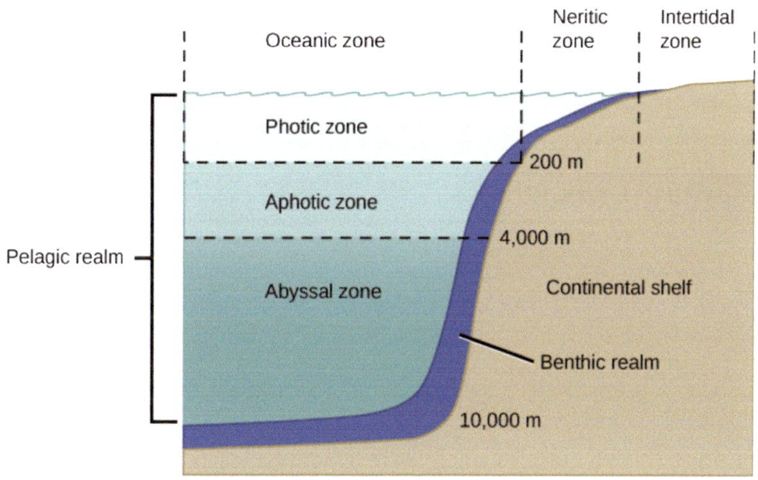

Figure 27. The ocean is divided into different zones based on water depth and distance from the shoreline.

The ocean is categorized by several areas or zones (Figure 27). All of the ocean's open water is referred to as the pelagic zone. The benthic zone extends along the ocean bottom from the shoreline to the deepest parts of the ocean floor. Within the pelagic realm is the photic zone, which is the portion of the ocean that light can penetrate (approximately 200 m or 650 ft). At depths greater than 200 m, light cannot penetrate; thus, this is referred to as the aphotic zone. The majority of the ocean is aphotic and lacks sufficient light for photosynthesis. The deepest part of the ocean, the Challenger Deep (in the Mariana Trench, located in the western Pacific Ocean), is about 11,000 m (about 6.8 mi) deep. To give some perspective on the depth of this trench, the ocean is, on average, 4267 m. These zones are relevant to freshwater lakes as well.

Marine Biomes

The ocean is the largest marine biome. It is a continuous body of salt water that is relatively uniform in chemical composition; it is a weak solution of mineral salts and decayed biological matter. Within the ocean, coral reefs are a second kind of marine biome. Estuaries, coastal areas where salt water and fresh water mix, form a third unique marine biome.

Ocean

The physical diversity of the ocean is a significant influence on plants, animals, and other organisms. The ocean is categorized into different zones based on how far light reaches into the water. Each zone has a distinct group of species adapted to the biotic and abiotic conditions particular to that zone.

The **intertidal zone**, which is the zone between high and low tide, is the oceanic region that is closest to land. Generally, most people think of this portion of the ocean as a sandy beach. In some cases, the intertidal zone is indeed a sandy beach, but it can also be rocky or muddy. Organisms are exposed to air and sunlight at low tide and are underwater most of the time, especially during high tide. Therefore, living things that thrive in the intertidal zone are adapted to being dry for long periods of time. The shore of the intertidal zone is also repeatedly struck by waves, and the organisms found there are adapted to withstand damage from the pounding action of the waves. The exoskeletons of shoreline crustaceans (such as the shore crab, *Carcinus maenas*) are tough and protect them from desiccation (drying out) and wave damage. Another consequence of the pounding waves is that few algae and plants establish themselves in the constantly moving rocks, sand, or mud.

The neritic zone extends from the intertidal zone to depths of about 200 m (or 650 ft) at the edge of the continental shelf. Because light can penetrate this depth, photosynthesis can occur. The water here contains silt and is well-oxygenated, low in pressure, and stable in temperature. Phytoplankton and floating Sargassum (a type of free-floating marine seaweed) provide a habitat for some sea life found in the neritic zone. Zooplankton, protists, small fishes, and shrimp are found in the neritic zone and are the base of the food chain for most of the world's fisheries.

Figure 28. Sea urchins, mussel shells, and starfish are often found in the intertidal zone, shown here in Kachemak Bay, Alaska. (credit: NOAA)

Beyond the neritic zone is the open ocean area known as the oceanic zone (Figure 1). Within the oceanic zone there is thermal stratification where warm and cold waters mix because of ocean currents. Abundant plankton serve as the base of the food chain for larger animals such as whales and dolphins. Nutrients are scarce and this is a relatively less productive part of the marine biome. When photosynthetic organisms and the protists and animals that feed on them die, their bodies fall to the bottom of the ocean where they remain. The majority of organisms in the aphotic zone include sea cucumbers (phylum Echinodermata) and other organisms that survive on the nutrients contained in the dead bodies of organisms in the photic zone. The deepest part of the ocean is the abyssal zone, which is at depths of 4000 m or greater. The abyssal zone is very cold and has very high pressure, high oxygen content, and low nutrient content. There are a variety of invertebrates and fishes found in this zone, but the abyssal zone does not have plants because of the lack of light. Cracks in the Earth's crust called hydrothermal vents are found primarily in the abyssal zone. Around these vents chemosynthetic bacteria utilize the hydrogen sulfide and other minerals emitted as an energy source and serve as the base of the food chain found in the abyssal zone. Beneath the water is the **benthic zone** , which is comprised of sand, silt, and dead organisms. This is a nutrient-rich portion of the ocean because of the dead organisms that fall from the upper layers of the ocean. Because of this high level of nutrients, a diversity of sponges, sea anemones, marine worms, sea stars, fishes, and bacteria exist.

Coral Reefs: An Overview

Figure 29. Coral reefs are formed by the calcium carbonate skeletons of coral organisms, which are marine invertebrates in the phylum Cnidaria. (credit: Terry Hughes)

Coral reefs are characterized by high biodiversity and the structures created by invertebrates that live in warm, shallow waters within the photic zone of the ocean. They are mostly found within 30 degrees north and south of the equator. The Great Barrier Reef is a well-known reef system located several miles off the northeastern coast of Australia. The coral organisms (members of phylum Cnidaria) are colonies of saltwater polyps that secrete a calcium carbonate skeleton. These calcium-rich skeletons slowly accumulate, forming the underwater reef (Figure 3). Corals found in shallower waters (at a depth of approximately 60 m or about 200 ft) have a mutualistic relationship with photosynthetic unicellular algae. The relationship provides corals with the majority of the nutrition and the energy they require. The waters in which these corals live are nutritionally poor and, without this mutualism, it would not be possible for large corals to grow. Some corals living in deeper and colder water do not have a mutualistic relationship with algae; these corals attain energy and nutrients using stinging cells on their tentacles to capture prey. It is estimated that more than 4,000 fish species inhabit coral reefs. These fishes can feed on coral, other invertebrates, or the seaweed and algae that are associated with the coral.

openstax COLLEGE™

Link to Learning: Watch this National Oceanic and Atmospheric Administration (NOAA) video to see marine ecologist Dr. Peter Etnoyer discusses his research on coral organisms.

3.5 EVOLUTION CONNECTION: GLOBAL DECLINE OF CORAL REEFS

It takes a long time to build a coral reef. The animals that create coral reefs have evolved over millions of years, continuing to slowly deposit the calcium carbonate that forms their characteristic ocean homes. Bathed in warm tropical waters, the coral animals and their symbiotic algal partners evolved to survive at the upper limit of ocean water temperature.

Together, climate change and human activity pose dual threats to the long-term survival of the world's coral reefs. As global warming due to fossil fuel emissions raises ocean temperatures, coral reefs are suffering. The excessive warmth causes the reefs to expel their symbiotic, food-producing algae, resulting in a phenomenon known as bleaching. When bleaching occurs, the reefs lose much of their characteristic color as the algae and the coral animals die if loss of the symbiotic zooxanthellae is prolonged.

Rising levels of atmospheric carbon dioxide further threaten the corals in other ways; as CO_2 dissolves in ocean waters, it lowers the pH and increases ocean acidity. As acidity increases, it interferes with the calcification that normally occurs as coral animals build their calcium carbonate homes. When a coral reef begins to die, species diversity plummets as animals lose food and shelter. Coral reefs are also economically important tourist destinations, so the decline of coral reefs poses a serious threat to coastal economies. Human population growth has damaged corals in other ways, too. As human coastal populations increase, the runoff of sediment and agricultural chemicals has increased, too, causing some of the once-clear tropical waters to become cloudy. At the same time, overfishing of popular fish species has allowed the predator species that eat corals to go unchecked.

Although a rise in global temperatures of 1–2°C (a conservative scientific projection) in the coming decades may not seem large, it is very significant to this biome. When change occurs rapidly, species can become extinct before evolution leads to new adaptations. Many scientists believe that global warming, with its rapid (in terms of evolutionary time) and inexorable increases in temperature, is tipping the balance beyond the point at which many of the world's coral reefs can recover.

Estuaries: An estuary is an area where a freshwater river or stream meets the ocean.

Estuaries are biomes that occur where a source of fresh water, such as a river, meets the ocean. Therefore, both fresh water and salt water are found in the same vicinity; mixing results in a diluted (brackish) saltwater. Estuaries form protected areas where many of the young offspring of crustaceans, mollusks, and fish begin their lives. Salinity is a very important factor that influences the organisms and the adaptations of the organisms found in estuaries. The salinity of estuaries varies and is based on the rate of flow of its freshwater sources. Once or twice a day, high tides bring salt water into the estuary. Low tides occurring at the same frequency reverse the current of salt water.

The short-term and rapid variation in salinity due to the mixing of fresh water and salt water is a difficult physiological challenge for the plants and animals that inhabit estuaries. Many estuarine plant species are halophytes: plants that can tolerate salty conditions. Halophytic plants are adapted to deal with the salinity resulting from saltwater on their roots or from sea spray. In some halophytes, filters in the roots remove the salt from the water that the plant absorbs. Other plants are able to pump oxygen into their roots. Animals, such as mussels and clams (phylum Mollusca), have developed behavioral adaptations that expend a lot of energy to function in this rapidly changing environment. When these animals are exposed to low salinity, they stop feeding, close their shells, and switch from aerobic respiration (in which they use gills) to anaerobic respiration (a process that does not require oxygen). When high tide returns to the estuary, the salinity and oxygen content of the water increases, and these animals open their shells, begin feeding, and return to aerobic respiration.

Freshwater Biomes: An Overview

Freshwater biomes include lakes and ponds (standing water) as well as rivers and streams (flowing water). They also include wetlands, which will be discussed later. Humans rely on freshwater biomes to provide aquatic resources for drinking water, crop irrigation, sanitation, and industry. These various roles and human benefits are referred to as ecosystem services. Lakes and ponds are found in terrestrial landscapes and are, therefore, connected with abiotic and biotic factors influencing these terrestrial biomes.

Lakes and Ponds: An Overview

Lakes and ponds can range in area from a few square meters to thousands of square kilometers. Temperature is an important abiotic factor affecting living things found in lakes and ponds. In the summer, thermal stratification of lakes and ponds occurs when the upper layer of water is warmed by the sun and does not mix with deeper, cooler water. Light can penetrate within the photic zone of the lake or pond. **Phytoplankton** (small photosynthetic organisms such as algae and cyanobacteria that float in the water) are found here and carry out photosynthesis, providing the base of the food web of lakes and ponds. **Zooplankton** (very small animals that float in the water), such as rotifers and small crustaceans, consume these phytoplankton. At the bottom of lakes and ponds, bacteria in the aphotic zone break down dead organisms that sink to the bottom.

Nitrogen and phosphorus are important limiting nutrients in lakes and ponds. Because of this, they are determining factors in the amount of phytoplankton growth in lakes and ponds. When there is a large input of nitrogen and phosphorus (from sewage and runoff from fertilized lawns and farms, for example), the growth of algae skyrockets, resulting in a large accumulation of algae called an algal bloom. Algal blooms (Figure 30) can become so extensive that they reduce light penetration in water. As a result, the lake or pond becomes aphotic and photosynthetic plants rooted in the lake bottom cannot survive. When the algae die and decompose, severe oxygen depletion of the water occurs. Fishes and other organisms that require oxygen are then more likely to die, and resulting dead zones are found across the globe. Lake Erie and the Gulf of Mexico represent freshwater and marine habitats where phosphorus control and storm water runoff pose significant environmental challenges.

Figure 30. The uncontrolled growth of algae in this lake has resulted in an algal bloom. (credit: Jeremy Nettleton)

3.6 RIVERS AND STREAMS: AN OVERVIEW

Rivers and streams are continuously moving bodies of water that carry large amounts of water from the source, or headwater, to a lake or ocean. The largest rivers include the Nile River in Africa, the Amazon River in South America, and the Mississippi River in North America.

Abiotic features of rivers and streams vary along the length of the river or stream. Streams begin at a point of origin referred to as source water. The source water is usually cold, low in nutrients, and clear. The channel (the width of the river or stream) is narrower than at any other place along the length of the river or stream. Because of this, the current is often faster here than at any other point of the river or stream.

The fast-moving water results in minimal silt accumulation at the bottom of the river or stream, therefore the water is clear. Photosynthesis here is mostly attributed to algae that are growing on rocks; the swift current inhibits the growth of phytoplankton. An additional input of energy can come from leaves or other organic material that falls into the river or stream from trees and other plants that border the water. When the leaves decompose, the organic material and nutrients in the leaves are returned to the water. Plants and animals have adapted to this fast-moving water. For instance, leeches (phylum Annelida) have elongated bodies and suckers on both ends. These suckers attach to the substrate, keeping the leech anchored in place. Freshwater trout species (phylum Chordata) are an important predator in these fast-moving rivers and streams.

As the river or stream flows away from the source, the width of the channel gradually widens and the current slows. This slow-moving water, caused by the gradient decrease and the volume increase as tributaries unite, has more sedimentation. Phytoplankton can also be suspended in slow-moving water. Therefore, the water will not be as clear as it is near the source. The water is also warmer. Worms (phylum Annelida) and insects (phylum Arthropoda) can be found burrowing into the mud. The higher order predator vertebrates (phylum Chordata) include waterfowl, frogs, and fishes. These predators must find food in these slow moving, sometimes murky, waters and, unlike the trout in the waters at the source, these vertebrates may not be able to use vision as their primary sense to find food. Instead, they are more likely to use taste or chemical cues to find prey.

Wetlands: An Overview

Figure 31. Located in southern Florida, Everglades National Park is vast array of wetland environments, including sawgrass marshes, cypress swamps, and estuarine mangrove forests. Here, a great egret walks among cypress trees. (credit: NPS)

Wetlands are environments in which the soil is either permanently or periodically saturated with water. Wetlands are different from lakes because wetlands are shallow bodies of water that may periodically dry out. Emergent vegetation consists of wetland plants that are rooted in the soil but have portions of leaves, stems, and flowers extending above the water's surface. There are several types of wetlands including marshes, swamps, bogs, mudflats, and salt marshes (Figure 31).

AT A GLANCE

Ecosystems exist underground, on land, at sea, and in the air. Organisms in an ecosystem acquire energy in a variety of ways, which is transferred between trophic levels as the energy flows from the base to the top of the food web, with energy being lost at each transfer. Mineral nutrients are cycled through ecosystems and their environment. Of particular importance are water, carbon, nitrogen, phosphorus, and sulfur. All of these cycles have major impacts on ecosystem structure and function. Ecosystems have been damaged by a variety of human activities that alter the natural biogeochemical cycles due to pollution, oil spills, and events causing global climate change. The health of the biosphere depends on understanding these cycles and how to protect the environment from irreversible damage. Earth has terrestrial and aquatic biomes. There are eight major terrestrial biomes: tropical rainforests, savannas, subtropical deserts, chaparral, temperate grasslands, temperate forests, boreal forests, and Arctic tundra. Temperature and precipitation, and variations in both, are key abiotic factors that shape the composition of animal and plant communities in terrestrial biomes. Sunlight is an important factor in bodies of water, especially those that are very deep, because of the role of photosynthesis in sustaining certain organisms. Other important factors include temperature, water movement, and salt content. Aquatic biomes include both freshwater and marine environments. Like terrestrial biomes, aquatic biomes are influenced by abiotic factors. In the case of aquatic biomes the abiotic factors include light, temperature, flow regime, and dissolved solids.

CHAPTER-4

POPULATION, COMMUNITIES AND ECOSYSTEMS

4.1 CONCEPTS OF POPULATION DYNAMICS AND DEMOGRAPHY

Imagine sailing down a river in a small motorboat on a weekend afternoon; the water is smooth, and you are enjoying the sunshine and cool breeze when suddenly you are hit in the head by a 20-pound silver carp. This is a risk now on many rivers and canal systems in Illinois and Missouri because of the presence of Asian carp. This fish—actually a group of species including the silver, black, grass, and big head carp—has been farmed and eaten in China for over 1,000 years. It is one of the most important aquaculture food resources worldwide. In the United States, however, Asian carp is considered a dangerous invasive species that disrupts ecological community structure to the point of threatening native species. The effects of invasive species (such as the Asian carp, kudzu vine, predatory snakehead fish, and zebra mussel) are just one aspect of what ecologists study to understand how populations interact within ecological communities, and what impact natural and human-induced disturbances have on the characteristics of communities.

Populations are dynamic entities. Their size and composition fluctuate in response to numerous factors, including seasonal and yearly changes in the environment, natural disasters such as forest fires and volcanic eruptions, and competition for resources between and within species. The study of populations is called **demography.**

Population Size and Density

Populations are characterized by their population size (total number of individuals) and their population density (number of individuals per unit area). A population may have a large number of individuals that are distributed densely, or sparsely. There are also populations with small numbers of individuals that may be dense or very sparsely distributed in a local area. Population size can affect potential for adaptation because it affects the amount of genetic variation present in the population. Density can have effects on interactions within a population such as competition for food and the ability of individuals to find a mate. Smaller organisms tend to be more densely distributed than larger organisms (Figure 1).

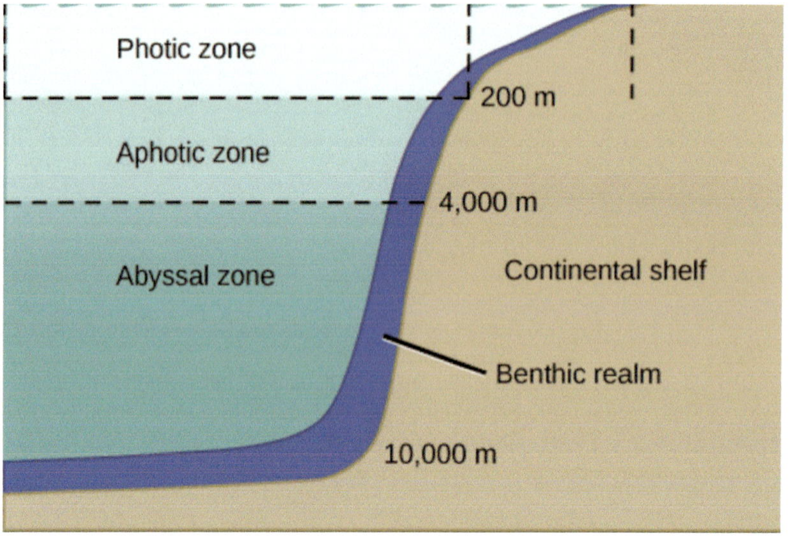

Figure 1. Australian mammals show a typical inverse relationship between population density and body size. As this graph shows, population density typically decreases with increasing body size. Why do you think this is the case?

Population Size Prediction

The most accurate way to determine population size is to count all of the individuals within the area. However, this method is usually not logistically or economically feasible, especially when studying large areas. Thus, scientists usually study populations by sampling a representative portion of each habitat and use this sample to make inferences about the population as a whole. The methods used to sample populations to determine their size and density are typically tailored to the characteristics of the organism being studied. For immobile organisms such as plants, or for very small and slow-moving organisms, a quadrat may be used. A **quadrat** is a square structure that is randomly located on the ground and used to count the number of individuals that lie within its boundaries. To obtain an accurate count using this method, the square must be placed at random locations within the habitat enough times to produce an accurate estimate.

For smaller mobile organisms, such as mammals, a technique called **mark and recapture** is often used. This method involves marking captured animals in and releasing them back into the environment to mix with the rest of the population. Later, a new sample is captured and scientists determine how many of the marked animals are in the new sample. This method assumes that the larger the population, the lower the percentage of marked organisms that will be recaptured since they will have mixed with more unmarked individuals. For example, if 80 field mice are captured, marked, and released into the forest, then a second trapping 100 field mice are captured and 20 of them are marked, the population size (*N*) can be determined using the following equation:

N = (number marked first catch x total number of second catch)/number marked second catch

Using our example, the equation would be:

$(80 \times 100)/20 = 400$

These results give us an estimate of 400 total individuals in the original population. The true number usually will be a bit different from this because of chance errors and possible bias caused by the sampling methods.

Species Distribution

In addition to measuring size and density, further information about a population can be obtained by looking at the distribution of the individuals throughout their range. A species distribution pattern is the distribution of individuals within a habitat at a particular point in time—broad categories of patterns are used to describe them.

Individuals within a population can be distributed at random, in groups, or equally spaced apart (more or less). These are known as **random, clumped, and uniform distribution patterns**, respectively (Figure 2). Different distributions reflect important aspects of the biology of the species. They also affect the mathematical methods required to estimate population sizes. An example of random distribution occurs with dandelion and other plants that have wind-dispersed seeds that germinate wherever they happen to fall in favorable environments. A clumped distribution, may be seen in plants that drop their seeds straight to the ground, such as oak trees; it can also be seen in animals that live in social groups (schools of fish or herds of elephants). Uniform distribution is observed in plants that secrete substances inhibiting the growth of nearby individuals (such as the release of toxic chemicals by sage plants). It is also seen in territorial animal species, such as penguins that maintain a defined territory for nesting. The territorial defensive behaviors of each individual create a regular pattern of distribution of similar-sized territories and individuals within those territories. Thus, the distribution of the individuals within a population provides more information about how they interact with each other than does a simple density measurement. Just as lower density species might have more difficulty finding a mate, solitary species with a random distribution might have a similar difficulty when compared to social species clumped together in groups.

Figure 2. Species may have a random, clumped, or uniform distribution. Plants such as (a) dandelions with wind-dispersed seeds tend to be randomly distributed. Animals such as (b) elephants that travel in groups exhibit a clumped distribution. Territorial birds such as (c) penguins tend to have a uniform distribution. (credit a: modification of work by Rosendahl; credit b: modification of work by Rebecca Wood; credit c: modification of work by Ben Tubby)

Life tables provide important information about the life history of an organism and the life expectancy of individuals at each age. They are modeled after actuarial tables used by the insurance industry for estimating human life expectancy. Life tables may include the probability of each age group dying before their next birthday, the percentage of surviving individuals dying at a particular age interval (their mortality rate, and their life expectancy at each interval. An example of a life table is shown in Table 1 from a study of Dall mountain sheep, a species native to northwestern North America. Notice that the population is divided into age intervals (column A).

As can be seen from the mortality rate data (column D), a high death rate occurred when the sheep were between six months and a year old, and then increased even more from 8 to 12 years old, after which there were few survivors. The data indicate that if a sheep in this population were to survive to age one, it could be expected to live another 7.7 years on average, as shown by the life-expectancy numbers in column E.

Another tool used by population ecologists is a **survivorship curve**, which is a graph of the number of individuals surviving at each age interval versus time. These curves allow us to compare the life histories of different populations (Figure 3). There are three types of survivorship curves. In a type I curve, mortality is low in the early and middle years and occurs mostly in older individuals. Organisms exhibiting a type I survivorship typically produce few offspring and provide good care to the offspring increasing the likelihood of their survival. Humans and most mammals exhibit a type I survivorship curve. In type II curves, mortality is relatively constant throughout the entire life span, and mortality is equally likely to occur at any point in the life span. Many bird populations provide examples of an intermediate or type II survivorship curve. In type III survivorship curves, early ages experience the highest mortality with much lower mortality rates for organisms that make it to advanced years. Type III organisms typically produce large numbers of offspring, but provide very little or no care for them. Trees and marine invertebrates exhibit a type III survivorship curve because very few of these organisms survive their younger years, but those that do make it to an old age are more likely to survive for a relatively long period of time.

Table 1. This life table of *Ovis dalli* shows the number of deaths, number of survivors, mortality rate, and life expectancy at each age interval for Dall mountain sheep.

Age interval (years)	Number dying in age interval out of 1000 born	Number surviving at beginning of age interval out of 1000 born	Mortality rate per 1000 alive at beginning of age interval	Life expectancy or mean lifetime remaining to those attaining age interval
0–0.5	54	1000	54.0	7.06
0.5–1	145	946	153.3	—
1–2	12	801	15.0	7.7
2–3	13	789	16.5	6.8
3–4	12	776	15.5	5.9
4–5	30	764	39.3	5.0
5–6	46	734	62.7	4.2
6–7	48	688	69.8	3.4
7–8	69	640	107.8	2.6
8–9	132	571	231.2	1.9
9–10	187	439	426.0	1.3
10–11	156	252	619.0	0.9
11–12	90	96	937.5	0.6
12–13	3	6	500.0	1.2
13–14	3	3	1000	0.7

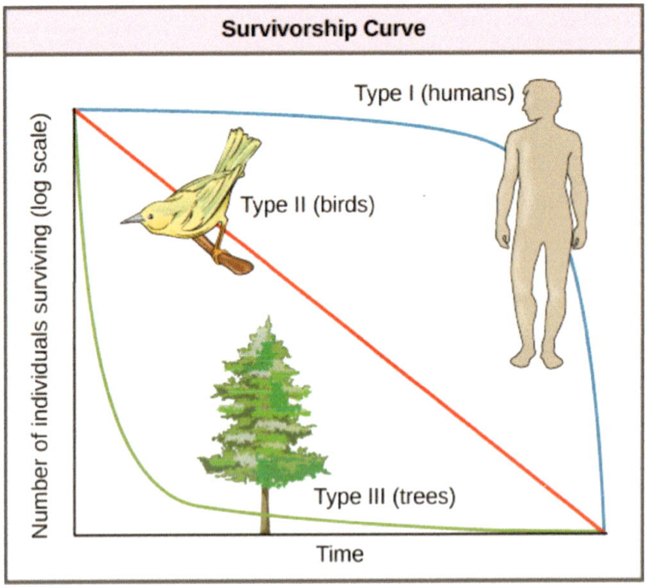

Figure 3. Survivorship curves show the distribution of individuals in a population according to age. Humans and most mammals have a Type I survivorship curve, because death primarily occurs in the older years. Birds have a Type II survivorship curve, as death at any age is equally probable. Trees have a Type III survivorship curve because very few survive the younger years, but after a certain age, individuals are much more likely to survive.

4.2 TYPES OF POPULATION GROWTH AND REGULATION

Population ecologists make use of a variety of methods to model population dynamics. An accurate model should be able to describe the changes occurring in a population and predict future changes.

Population Growth

The two simplest models of population growth use deterministic equations (equations that do not account for random events) to describe the rate of change in the size of a population over time. The first of these models, **exponential growth**, describes populations that increase in numbers without any limits to their growth. The second model, **logistic growth**, introduces limits to reproductive growth that become more intense as the population size increases. Neither model adequately describes natural populations, but they provide points of comparison.

Exponential Growth

Charles Darwin, in developing his theory of natural selection, was influenced by the English clergyman Thomas Malthus. Malthus published his book in 1798 stating that populations with abundant natural resources grow very rapidly. However, they limit further growth by depleting their resources. The early pattern of accelerating population size is called exponential growth (Figure 1).

The best example of exponential growth in organisms is seen in bacteria. Bacteria are prokaryotes that reproduce quickly, about an hour for many species. If 1000 bacteria are placed in a large flask with an abundant supply of nutrients (so the nutrients will not become quickly depleted), the number of bacteria will have doubled from 1000 to 2000 after just an hour. In another hour, each of the 2000 bacteria will divide, producing 4000 bacteria. After the third hour, there should be 8000 bacteria in the flask. The important concept of exponential growth is that the growth rate—the number of organisms added in each reproductive generation—is itself increasing; that is, the population size is increasing at a greater and greater rate. After 24 of these cycles, the population would have increased from 1000 to more than 16 billion bacteria. When the population size, N, is plotted over time, a J-shaped growth curve is produced (Figure 1).

The bacteria-in-a-flask example is not truly representative of the real world where resources are usually limited. However, when a species is introduced into a new habitat that it finds suitable, it may show exponential growth for a while. In the case of the bacteria in the flask, some bacteria will die during the experiment and thus not reproduce; therefore, the growth rate is lowered from a maximal rate in which there is no mortality.

Logistic Growth

Extended exponential growth is possible only when infinite natural resources are available; this is not the case in the real world. Charles Darwin recognized this fact in his description of the "struggle for existence," which states that individuals will compete, with members of their own or other species, for limited resources. The successful ones are more likely to survive and pass on the traits that made them successful to the next generation at a greater rate (natural selection). To model the reality of limited resources, population ecologists developed the logistic growth model.

In the real world, with its limited resources, exponential growth cannot continue indefinitely. Exponential growth may occur in environments where there are few individuals and plentiful resources, but when the number of individuals gets large enough, resources will be depleted and the growth rate will slow down. Eventually, the growth rate will plateau or level off (Figure 1). This population size, which is determined by the maximum population size that a particular environment can sustain, is called the **carrying capacity**, symbolized as K. In real populations, a growing population often overshoots its carrying capacity and the death rate increases beyond the birth rate causing the population size to decline back to the carrying capacity or below it. Most populations usually fluctuate around the carrying capacity in an undulating fashion rather than existing right at it.

Carrying Capacity and the Logistic Model

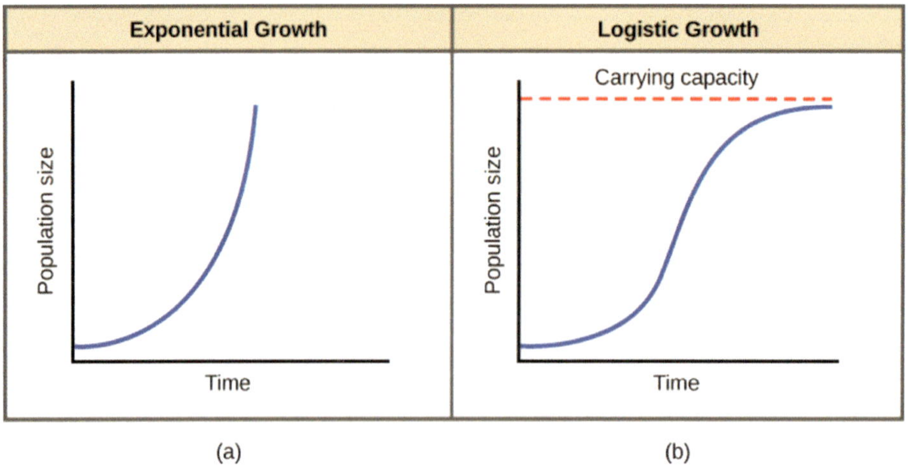

Figure 4. When resources are unlimited, populations exhibit (a) exponential growth, shown in a J-shaped curve. When resources are limited, populations exhibit (b) logistic growth. In logistic growth, population expansion decreases as resources become scarce, and it levels off when the carrying capacity of the environment is reached. The logistic growth curve is S-shaped.

A graph of logistic growth yields the S-shaped curve (Figure 1). It is a more realistic model of population growth than exponential growth. There are three different sections to an S-shaped curve. Initially, growth is exponential because there are few individuals and ample resources available. Then, as resources begin to become limited, the growth rate decreases. Finally, the growth rate levels off at the carrying capacity of the environment, with little change in population number over time.

Logistic Growth: An Example

Yeast, a unicellular fungus used to make bread and alcoholic beverages, exhibits the classical S-shaped curve when grown in a test tube (Figure 5a). Its growth levels off as the population depletes the nutrients that are necessary for its growth. In the real world, however, there are variations to this idealized curve. Examples in wild populations include sheep and harbor seals (Figure 5b). In both examples, the population size exceeds the carrying capacity for short periods of time and then falls below the carrying capacity afterwards. This fluctuation in population size continues to occur as the population oscillates around its carrying capacity. Still, even with this oscillation the logistic model is confirmed.

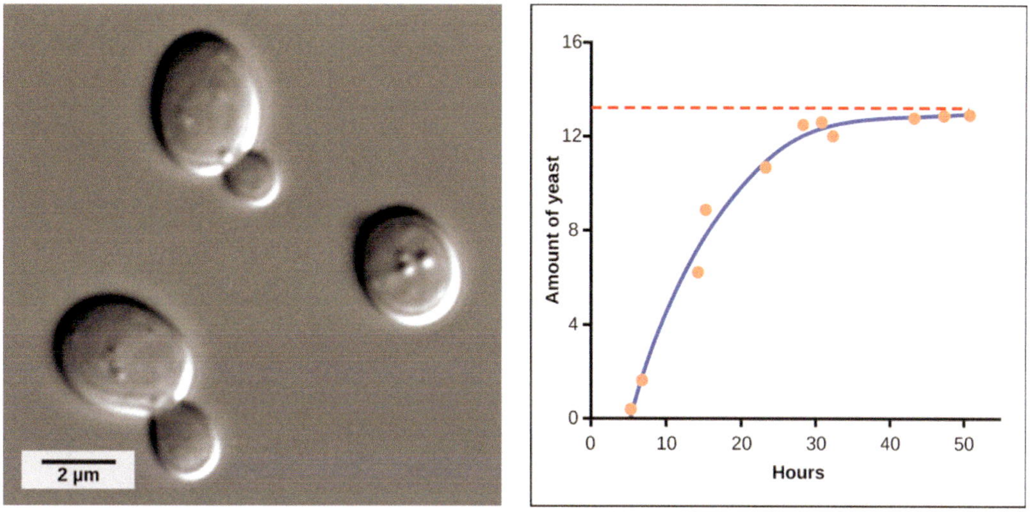

(a)

(b)

Figure 5. (a) Yeast grown in ideal conditions in a test tube shows a classical S-shaped logistic growth curve, whereas (b) a natural population of seals shows real-world fluctuation. The yeast is visualized using differential interference contrast light micrography. (credit a: scale-bar data from Matt Russell)

Population Dynamics and Model of Population Regulation

The logistic model of population growth, while valid in many natural populations and a useful model, is a simplification of real-world population dynamics. Implicit in the model is that the carrying capacity of the environment does not change, which is not the case. The carrying capacity varies annually. For example, some summers are hot and dry whereas others are cold and wet; in many areas, the carrying capacity during the winter is much lower than it is during the summer. Also, natural events such as earthquakes, volcanoes, and fires can alter an environment and hence its carrying capacity. Additionally, populations do not usually exist in isolation. They share the environment with other species, competing with them for the same resources (interspecific competition). These factors are also important to understanding how a specific population will grow.

Why Did the Woolly Mammoth Go Extinct?

(a)

(b)

(c)

Figure 6. The three images include: (a) 1916 mural of a mammoth herd from the American Museum of Natural History, (b) the only stuffed mammoth in the world is in the Museum of Zoology located in St. Petersburg, Russia, and (c) a one-month-old baby mammoth, named Lyuba, discovered in Siberia in 2007. (credit a: modification of work by Charles R. Knight; credit b: modification of work by "Tanapon"/Flickr; credit c: modification of work by Matt Howry)

Most populations of woolly mammoths went extinct about 10,000 years ago, soon after paleontologists believe humans began to colonize North America and northern Eurasia (Figure 6). A mammoth population survived on Wrangel Island, in the East Siberian Sea, and was isolated from human contact until as recently as 1700 BC. We know a lot about these animals from carcasses found frozen in the ice of Siberia and other northern regions.

It is commonly thought that climate change and human hunting led to their extinction. A 2008 study estimated that climate change reduced the mammoth's range from 3,000,000 square miles 42,000 years ago to 310,000 square miles 6,000 years ago.2 Through archaeological evidence of kill sites, it is also well documented that humans hunted these animals. A 2012 study concluded that no single factor was exclusively responsible for the extinction of these magnificent creatures.3 In addition to climate change and reduction of habitat, scientists demonstrated another important factor in the mammoth's extinction was the migration of human hunters across the Bering Strait to North America during the last ice age 20,000 years ago.

The maintenance of stable populations was and is very complex, with many interacting factors determining the outcome. It is important to remember that humans are also part of nature. Once we contributed to a species' decline using primitive hunting technology only.

Models of Population Based on Demographics

Population ecologists have hypothesized that suites of characteristics may evolve in species that lead to particular adaptations to their environments. These adaptations impact the kind of population growth their species experience. Life history characteristics such as birth rates, age at first reproduction, the numbers of offspring, and even death rates evolve just like anatomy or behavior, leading to adaptations that affect population growth. Population ecologists have described a continuum of life-history "strategies" with K-selected species on one end and r-selected species on the other. K-selected species are adapted to stable, predictable environments. Populations of K-selected species tend to exist close to their carrying capacity. These species tend to have larger, but fewer, offspring and contribute large amounts of resources to each offspring. Elephants would be an example of a K-selected species. r-selected species are adapted to unstable and unpredictable environments. They have large numbers of small offspring. Animals that are r-selected do not provide a lot of resources or parental care to offspring, and the offspring are relatively self-sufficient at birth. Examples of r-selected species are marine invertebrates such as jellyfish and plants such as the dandelion. The two extreme strategies are at two ends of a continuum on which real species life histories will exist. In addition, life history strategies do not need to evolve as suites, but can evolve independently of each other, so each species may have some characteristics that trend toward one extreme or the other.

4.3 CONCEPTS OF HUMAN POPULATION

Concepts of animal population dynamics can be applied to human population growth. Humans are not unique in their ability to alter their environment. For example, beaver dams alter the stream environment where they are built. Humans, however, have the ability to alter their environment to increase its carrying capacity, sometimes to the detriment of other species. Earth's human population and their use of resources are growing rapidly, to the extent that some worry about the ability of Earth's environment to sustain its human population. Long-term exponential growth carries with it the potential risks of famine, disease, and large-scale death, as well as social consequences of crowding such as increased crime.

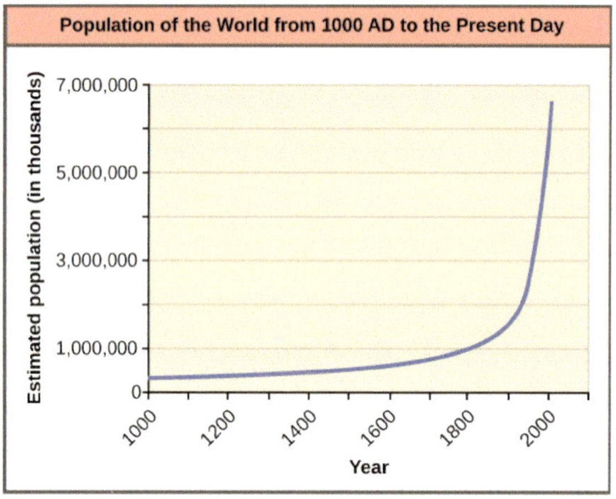

Figure 7. Human population growth since 1000 AD is exponential.

Human technology and particularly our harnessing of the energy contained in fossil fuels have caused unprecedented changes to Earth's environment, altering ecosystems to the point where some may be in danger of collapse. Changes on a global scale including depletion of the ozone layer, desertification and topsoil loss, and global climate change are caused by human activities.

Human population Growth

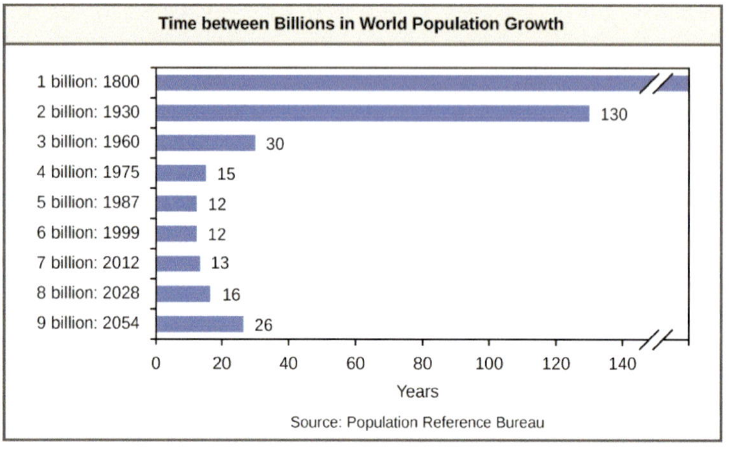

Figure 8. The time between the addition of each billion human beings to Earth decreases over time. (credit: modification of work by Ryan T. Cragun)

The fundamental cause of the acceleration of growth rate for humans in the past 200 years has been the reduced death rate due to changes in public health and sanitation. Clean drinking water and proper disposal sewage has drastically improved health in developed nations. Also, medical innovations such as the use of antibiotics and vaccines have decreased the ability of infectious disease to limit human population growth. In the past, diseases such as the bubonic plaque of the fourteenth century killed between 30 and 60 percent of Europe's population and reduced the overall world population by as many as one hundred million people. Naturally, infectious disease continues to have an impact on human population growth, especially in poorer nations. For example, life expectancy in sub-Saharan Africa, which was increasing from 1950 to 1990, began to decline after 1985 largely as a result of HIV/AIDS mortality. The reduction in life expectancy caused by HIV/AIDS was estimated to be 7 years for 2005.

openstax COLLEGE™

Link to Learning: Click through this interactive view of how human populations have changed over time.

Technological advances of the industrial age have also supported population growth through urbanization and advances in agriculture. These advances in technology were possible, in part, due to the exploitation of fossil fuels.

Age Structure, Population Growth, and Economic Development

The age structure of a population is an important factor in population dynamics. Age structure is the proportion of a population in different age classes. Models that incorporate age structure allow better prediction of population growth, plus the ability to associate this growth with the level of economic development in a region. Countries with rapid growth have a pyramidal shape in their age structure diagrams, showing a preponderance of younger individuals, many of whom are of reproductive age. This pattern is most often observed in underdeveloped countries where individuals do not live to old age because of less-than-optimal living conditions, and there is a high birth rate. Age structures of areas with slow growth, including developed countries such as the United States, still have a pyramidal structure, but with many fewer young and reproductive-aged individuals and a greater proportion of older individuals. Other developed countries, such as Italy, have zero population growth. The age structure of these populations is more conical, with an even greater percentage of middle-aged and older individuals. The actual growth rates in different countries are shown in Figure 9, with the highest rates tending to be in the less economically developed countries of Africa and Asia.

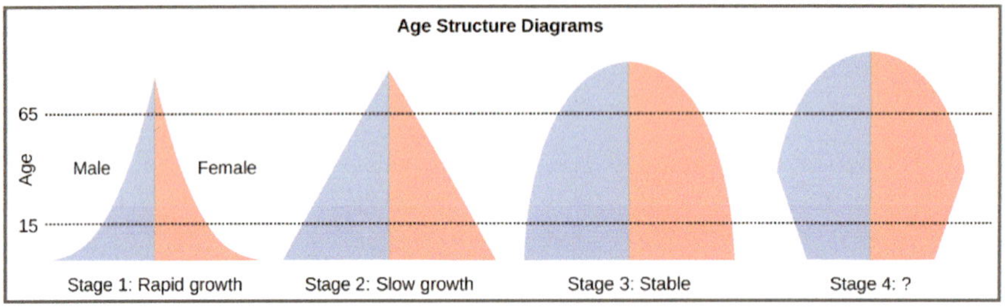

Figure 9. Typical age structure diagrams are shown. The rapid growth diagram narrows to a point, indicating that the number of individuals decreases rapidly with age. In the slow growth model, the number of individuals decreases steadily with age. Stable population diagrams are rounded on the top, showing that the number of individuals per age group decreases gradually, and then increases for the older part of the population.

Consequences of Human Population Growth in the Long Term

Many dire predictions have been made about the world's population leading to a major crisis called the "population explosion." In the 1968 book *The Population Bomb*, biologist Paul R. Ehrlich wrote, "The battle to feed all of humanity is over. In the 1970s hundreds of millions of people will starve to death in spite of any crash programs embarked upon now. At this late date nothing can prevent a substantial increase in the world death rate."

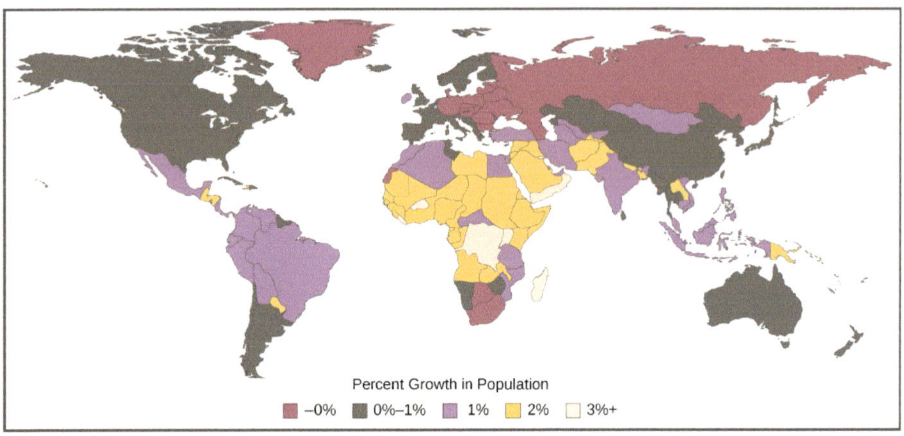

Figure 10. The percent growth rate of population in different countries is shown. Notice that the highest growth is occurring in less economically developed countries in Africa and Asia.

While these predictions obviously didn't bear fruit, the laws of exponential population growth are still in effect, and unchecked human population growth cannot continue indefinitely. Efforts to moderate population control led to the **one-child policy** in China, which imposes fines on urban couples who have more than one child. Due to the fact that some couples wish to have a male heir, many Chinese couples continue to have more than one child. The effectiveness of the policy in limiting overall population growth is controversial, as is the policy itself. Moreover, there are stories of female infanticide having occurred in some of the more rural areas of the country. Family planning education programs in other countries have had highly positive effects on limiting population growth rates and increasing standards of living.

Link to Learning: Visit this website and select "Launch the movie" for an animation discussing the global impacts of human population growth.

In spite of population control policies, the human population continues to grow. The United Nations estimates the future world population size to be 11.2 billion people by the year 2100. There is no way to know whether human population growth will moderate to the point where the crisis described by Dr. Ehrlich will be averted. Another consequence of population growth is the change and degradation of the natural environment. Many countries have attempted to reduce the human impact on climate change by limiting their emission of greenhouse gases. However, a global climate change treaty remains elusive, and many underdeveloped countries trying to improve their economic condition may be less likely to agree with such provisions without compensation if it means slowing their economic development. Furthermore, the role of human activity in causing climate change has become a hotly debated socio-political issue in some developed countries, including the United States, despite the overwhelming scientific evidence. Thus, we enter the future with considerable uncertainty about our ability to curb human population growth and protect our environment to maintain the carrying capacity for the human species.

4.4 CONCEPTS OF COMMUNITY ECOLOGY

Populations typically do not live in isolation from other species. Populations that interact within a given habitat form a **community**. The number of species occupying the same habitat and their relative abundance is known as the **diversity** of the community. Areas with low species diversity, such as the glaciers of Antarctica, still contain a wide variety of living organisms, whereas the diversity of tropical rainforests is so great that it cannot be accurately assessed. Scientists study ecology at the community level to understand how species interact with each other and compete for the same resources.

Predation and Herbivory

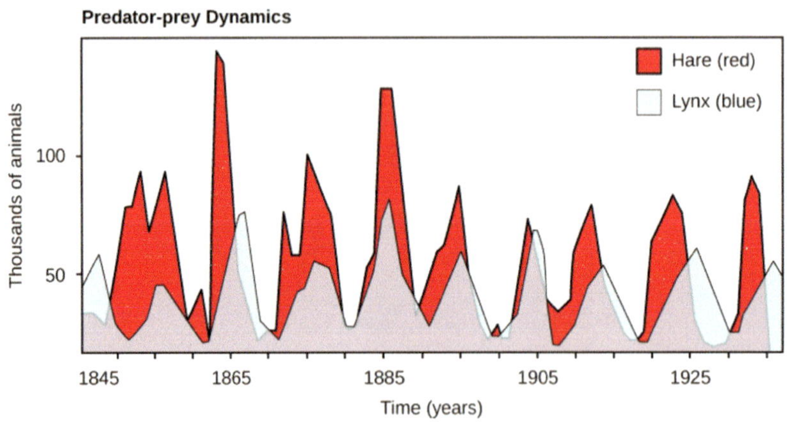

Figure 11. The cycling of snowshoe hare and lynx populations in Northern Ontario is an example of predator-prey dynamics.

Perhaps the classical example of species interaction is the predator-prey relationship. The narrowest definition of predation describes individuals of one population that kill and then consume the individuals of another population. Population sizes of predators and prey in a community are not constant over time, and they may vary in cycles that appear to be related. The most often cited example of predator-prey population dynamics is seen in the cycling of the lynx (predator) and the snowshoe hare (prey), using 100 years of trapping data from North America (Figure 11). This cycling of predator and prey population sizes has a period of approximately ten years, with the predator population lagging one to two years behind the prey population. An apparent explanation for this pattern is that as the hare numbers increase, there is more food available for the lynx, allowing the lynx population to increase as well. When the lynx population grows to a threshold level, however, they kill so many hares that hare numbers begin to decline, followed by a decline in the lynx population because of scarcity of food. When the lynx population is low, the hare population size begins to increase due, in part, to low predation pressure, starting the cycle anew.

Mechanisms of Defense against Predation and Herbivory

Predation and predator avoidance are strong influenced by natural selection. Any heritable character that allows an individual of a prey population to better evade its predators will be represented in greater numbers in later generations. Likewise, traits that allow a predator to more efficiently locate and capture its prey will lead to a greater number of offspring and an increase in the commonness of the trait within the population. Such ecological relationships between specific populations lead to adaptations that are driven by reciprocal evolutionary responses in those populations. Species have evolved numerous mechanisms to escape predation (including **herbivory**, the consumption of plants for food). Defenses may be mechanical, chemical, physical, or behavioral.

(a) (b)

Figure 12. The (a) honey locust tree uses thorns, a mechanical defense, against herbivores, while the (b) foxglove uses a chemical defense: toxins produces by the plant can cause nausea, vomiting, hallucinations, convulsions, or death when consumed. (credit a: modification of work by Huw Williams; credit b: modification of work by Philip Jägenstedt)

Mechanical defenses, such as the presence of armor in animals or thorns in plants, discourage predation and herbivory by discouraging physical contact (Figure 2a). Many animals produce or obtain chemical defenses from plants and store them to prevent predation. Many plant species produce secondary plant compounds that serve no function for the plant except that they are toxic to animals and discourage consumption. For example, the foxglove produces several compounds, including digitalis, that are extremely toxic when eaten (Figure 2b). (Biomedical scientists have repurposed the chemical produced by foxglove as a heart medication, which has saved lives for many decades.) Many species use their body shape and coloration to avoid being detected by predators. The tropical walking stick is an insect with the coloration and body shape of a twig, which makes it very hard to see when it is stationary against a background of real twigs (Figure 3a). In another example, the chameleon can change its color to match its surroundings (Figure 12b).

(a) (b)

Figure 13. (a) The tropical walking stick and (b) the chameleon use their body shape and/or coloration to prevent detection by predators. (credit a: modification of work by Linda Tanner; credit b: modification of work by Frank Vassen)

Figure 14. The fire-bellied toad has bright coloration on its belly that serves to warn potential predators that it is toxic. (credit: modification of work by Roberto Verzo)

Some species use coloration as a way of warning predators that they are distasteful or poisonous. For example, the monarch butterfly caterpillar sequesters poisons from its food (plants and milkweeds) to make itself poisonous or distasteful to potential predators. The caterpillar is bright yellow and black to advertise its toxicity. The caterpillar is also able to pass the sequestered toxins on to the adult monarch, which is also dramatically colored black and red as a warning to potential predators. Fire-bellied toads produce toxins that make them distasteful to their potential predators (Figure 14). They have bright red or orange coloration on their bellies, which they display to a potential predator to advertise their poisonous nature and discourage an attack. Warning coloration only works if a predator uses eyesight to locate prey and can learn—a naïve predator must experience the negative consequences of eating one before it will avoid other similarly colored individuals.

Figure 15. One form of mimicry is when a harmless species mimics the coloration of a harmful species, as is seen with the (a) wasp (Polistes sp.) and the (b) hoverfly (Syrphus sp.). (credit: modification of work by Tom Ings)

While some predators learn to avoid eating certain potential prey because of their coloration, other species have evolved mechanisms to mimic this coloration to avoid being eaten, even though they themselves may not be unpleasant to eat or contain toxic chemicals. In some cases of **mimicry**, a harmless species imitates the warning coloration of a harmful species. Assuming they share the same predators, this coloration then protects the harmless ones. Many insect species mimic the coloration of wasps, which are stinging, venomous insects, thereby discouraging predation (Figure 15).

Figure 16. Several unpleasant-tasting *Heliconius* butterfly species share a similar color pattern with better-tasting varieties, an example of mimicry. (credit: Joron M, Papa R, Beltrán M, Chamberlain N, Mavárez J, et al.)

In other cases of mimicry, multiple species share the same warning coloration, but all of them actually have defenses. The commonness of the signal improves the compliance of all the potential predators. Figure 16 shows a variety of foul-tasting butterflies with similar coloration.

Link to Learning: Go to this website to view stunning examples of mimicry.

Principle of Competitive Exclusion

Resources are often limited within a habitat and multiple species may compete to obtain them. Ecologists have come to understand that all species have an **ecological niche:** the unique set of resources used by a species, which includes its interactions with other species. The **competitive exclusion principle** states that two species cannot occupy the exact same niche in a habitat. In other words, different species cannot coexist in a community if they are competing for all the same resources. It is important to note that competition is bad for both competitors because it wastes energy. The competitive exclusion principle works because if there is competition between two species for the same resources, then natural selection will favor traits that lessen reliance on the shared resource, thus reducing competition. If either species is unable to evolve to reduce competition, then the species that most efficiently exploits the resource will drive the other species to extinction. An experimental example of this principle is shown in Figure 7 with two protozoan species: *Paramecium aurelia* and *Paramecium caudatum*. When grown individually in the laboratory, they both thrive. But when they are placed together in the same test tube (habitat), *P. aurelia* outcompetes *P. caudatum* for food, leading to the latter's eventual extinction.

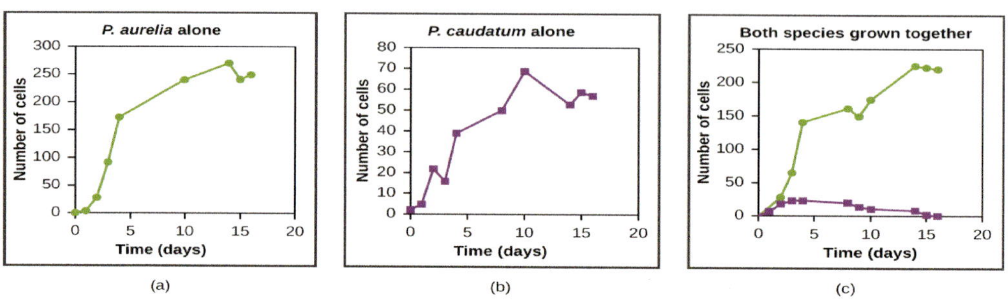

Figure 17. *Paramecium aurelia* and *Paramecium caudatum* grow well individually, but when they compete for the same resources, the P. aurelia outcompetes the P. caudatum.

Symbiosis

Symbiotic relationships are close, long-term interactions between individuals of different species. Symbioses may be commensal, in which one species benefits while the other is neither harmed nor benefited; mutualistic, in which both species benefit; or parasitic, in which the interaction harms one species and benefits the other.

Figure 18. The southern masked-weaver is starting to make a nest in a tree in Zambezi Valley, Zambia. This is an example of a commensal relationship, in which one species (the bird) benefits, while the other (the tree) neither benefits nor is harmed. (credit: "Hanay"/Wikimedia Commons)

Commensalism occurs when one species benefits from a close prolonged interaction, while the other neither benefits or is harmed. Birds nesting in trees provide an example of a commensal relationship (Figure 18). The tree is not harmed by the presence of the nest among its branches. The nests are light and produce little strain on the structural integrity of the branch, and most of the leaves, which the tree uses to get energy by photosynthesis, are above the nest so they are unaffected. The bird, on the other hand, benefits greatly. If the bird had to nest in the open, its eggs and young would be vulnerable to predators. Many potential commensal relationships are difficult to identify because it is difficult to prove that one partner does not derive some benefit from the presence of the other.

A second type of symbiotic relationship is called **mutualism**, in which two species benefit from their interaction. For example, termites have a mutualistic relationship with protists that live in the insect's gut (Figure 9a). The termite benefits from the ability of the protists to digest cellulose. However, the protists are able to digest cellulose only because of the presence of symbiotic bacteria within their cells that produce the cellulase enzyme. The termite itself cannot do this; without the protozoa, it would not be able to obtain energy from its food (cellulose from the wood it chews and eats). The protozoa benefit by having a protective environment and a constant supply of food from the wood chewing actions of the termite. In turn, the protists benefit from the enzymes provided by their bacterial endosymbionts, while the bacteria benefit from a doubly protective environment and a constant source of nutrients from two hosts. Lichen are a mutualistic relationship between a fungus and photosynthetic algae or cyanobacteria (Figure 9b). The glucose produced by the algae provides nourishment for both organisms, whereas the physical structure of the lichen protects the algae from the elements and makes certain nutrients in the atmosphere more available to the algae. The algae of lichens can live independently given the right environment, but many of the fungal partners are unable to live on their own.

A parasite is an organism that feeds off another without immediately killing the organism it is feeding on. In parasitism,the parasite benefits, but the organism being fed upon, the host, is harmed. The host is usually weakened by the parasite as it siphons resources the host would normally use to maintain itself. Parasites may kill their hosts, but there is usually selection to slow down this process to allow the parasite time to complete its reproductive cycle before it or its offspring are able to spread to another host. Parasitism is a form of predation.

(a) (b)

Figure 19. (a) Termites form a mutualistic relationship with symbiotic protozoa in their guts, which allow both organisms to obtain energy from the cellulose the termite consumes. (b) Lichen is a fungus that has symbiotic photosynthetic algae living in close association. (credit a: modification of work by Scott Bauer, USDA; credit b: modification of work by Cory Zanker)

The reproductive cycles of parasites are often very complex, sometimes requiring more than one host species. A tapeworm causes disease in humans when contaminated and under-cooked meat such as pork, fish, or beef is consumed (Figure 20). The tapeworm can live inside the intestine of the host for several years, benefiting from the host's food, and it may grow to be over 50 feet long by adding segments. The parasite moves from one host species to a second host species in order to complete its life cycle.

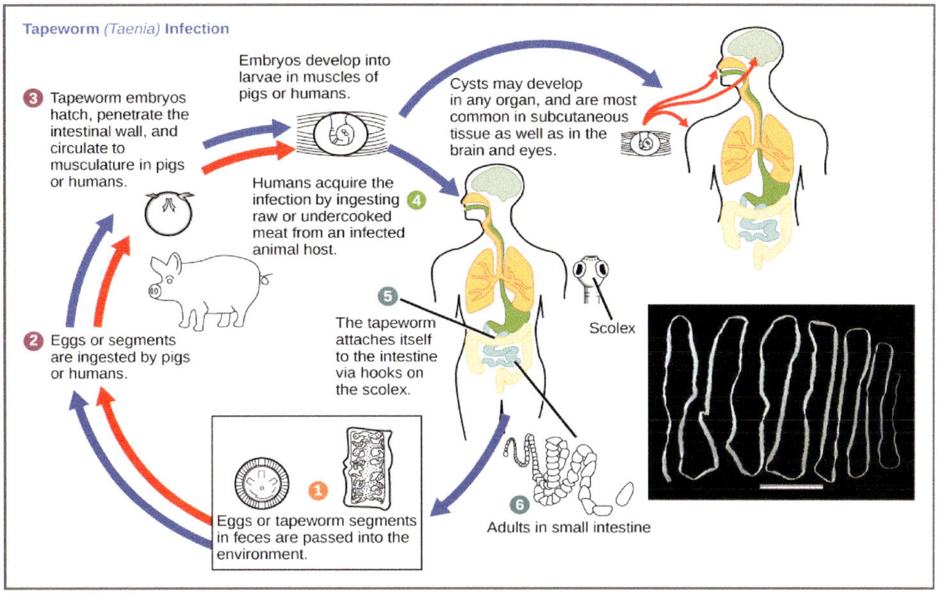

Figure 20. This diagram shows the life cycle of the tapeworm, a human worm parasite. (credit: modification of work by CDC)

Communities are Complex Systems

Communities are complex systems that can be characterized by their structure (the number and size of populations and their interactions) and dynamics (how the members and their interactions change over time). Understanding community structure and dynamics allows us to minimize impacts on ecosystems and manage ecological communities we benefit from.

Ecologists have extensively studied one of the fundamental characteristics of communities: **biodiversity**. One measure of biodiversity used by ecologists is the number of different species in a particular area and their relative abundance. The area in question could be a habitat, a biome, or the entire biosphere. **Species richness** is the term used to describe the number of species living in a habitat or other unit. Species richness varies across the globe (Figure 21). Species richness is related to latitude: the greatest species richness occurs near the equator and the lowest richness occurs near the poles. The exact reasons for this are not clearly understood. Other factors besides latitude influence species richness as well. For example, ecologists studying islands found that biodiversity varies with island size and distance from the mainland.

Relative abundance is the number individuals in a species relative to the total number of individuals in all species within a system. Foundation species, described below, often have the highest relative abundance of species.

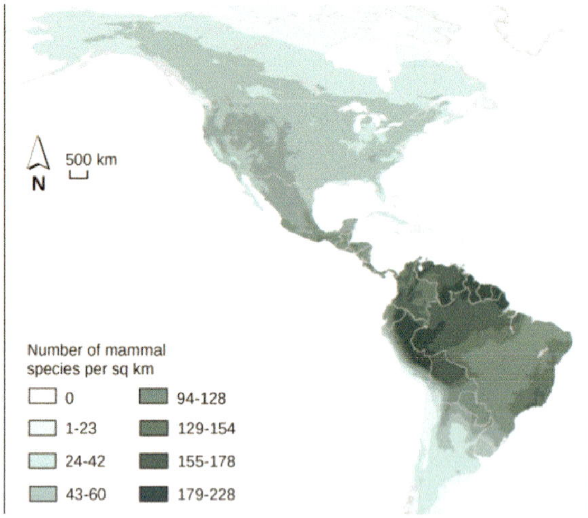

Figure 21. The greatest species richness for mammals in North America is associated in the equatorial latitudes. (credit: modification of work by NASA, CIESIN, Columbia University)

Foundation species are considered the "base" or "bedrock" of a community, having the greatest influence on its overall structure. They are often primary producers, and they are typically an abundant organism. For example, kelp, a species of brown algae, is a foundation species that forms the basis of the kelp forests off the coast of California.

Foundation species may physically modify the environment to produce and maintain habitats that benefit the other organisms that use them. Examples include the kelp described above or tree species found in a forest. The photosynthetic corals of the coral reef also provide structure by physically modifying the environment (Figure 22). The exoskeletons of living and dead coral make up most of the reef structure, which protects many other species from waves and ocean currents.

Figure 22. Coral is the foundation species of coral reef ecosystems. (credit: Jim E. Maragos, USFWS)

A **keystone species** is one whose presence has inordinate influence in maintaining the prevalence of various species in an ecosystem, the ecological community's structure, and sometimes its biodiversity. *Pisaster ochraceus*, the intertidal sea star, is a keystone species in the northwestern portion of the United States (Figure 13). Studies have shown that when this organism is removed from communities, mussel populations (their natural prey) increase, which completely alters the species composition and reduces biodiversity. Another keystone species is the banded tetra, a fish in tropical streams, which supplies nearly all of the phosphorus, a necessary inorganic nutrient, to the rest of the community. The banded tetra feeds largely on insects from the terrestrial ecosystem and then excretes phosphorus into the aquatic ecosystem. The relationships between populations in the community, and possibly the biodiversity, would change dramatically if these fish were to become extinct.

Figure 23. Sea stars are keystone species of the intertidal zone.

4.5 BIOLOGY IN ACTION

Invasive species are non-native organisms that, when introduced to an area out of its native range, alter the community they invade. In the United States, invasive species like the purple loosestrife (*Lythrum salicaria*) and the zebra mussel (*Dreissena polymorpha*) have drastically altered the ecosystems they invaded. Some well-known invasive animals include the emerald ash borer (*Agrilus planipennis*) and the European starling (*Sturnus vulgaris*). Whether enjoying a forest hike, taking a summer boat trip, or simply walking down an urban street, you have likely encountered an invasive species.

One of the many recent proliferations of an invasive species concerns the Asian carp in the United States. Asian carp were introduced to the United States in the 1970s by fisheries (commercial catfish ponds) and by sewage treatment facilities that used the fish's excellent filter feeding abilities to clean their ponds of excess plankton. Some of the fish escaped, and by the 1980s they had colonized many waterways of the Mississippi River basin, including the Illinois and Missouri Rivers.

Voracious feeders and rapid reproducers, Asian carp may outcompete native species for food and could lead to their extinction. One species, the grass carp, feeds on phytoplankton and aquatic plants. It competes with native species for these resources and alters nursery habitats for other fish by removing aquatic plants. In some parts of the Illinois River, Asian carp constitute 95 percent of the community's biomass. Although edible, the fish is bony and not desired in the United States. The Great Lakes and their prized salmon and lake trout fisheries are being threatened by Asian carp. The carp are not yet present in the Great Lakes, and attempts are being made to prevent its access to the lakes through the Chicago Ship and Sanitary Canal, which is the only connection between the Mississippi River and Great Lakes basins. To prevent the Asian carp from leaving the canal, a series of electric barriers have been used to discourage their migration; however, the threat is significant enough that several states and Canada have sued to have the Chicago channel permanently cut off from Lake Michigan. Local and national politicians have weighed in on how to solve the problem. In general, governments have been ineffective in preventing or slowing the introduction of invasive species.

Basic Concepts of Community Dynamics

Community dynamics are the changes in community structure and composition over time, often following environmental disturbances such as volcanoes, earthquakes, storms, fires, and climate change. Communities with a relatively constant number of species are said to be at equilibrium. The equilibrium is dynamic with species identities and relationships changing over time, but maintaining relatively constant numbers. Following a disturbance, the community may or may not return to the equilibrium state. Succession describes the sequential appearance and disappearance of species in a community over time after a severe disturbance. In primary succession, newly exposed or newly formed rock is colonized by living organisms. In secondary succession, a part of an ecosystem is disturbed and remnants of the previous community remain. In both cases, there is a sequential change in species until a more or less permanent community develops.

Primary Succession and Pioneer Species

Primary succession occurs when new land is formed, or when the soil and all life is removed from pre-existing land. An example of the former is the eruption of volcanoes on the Big Island of Hawaii, which results in lava that flows into the ocean and continually forms new land. From this process, approximately 32 acres of land are added to the Big Island each year. An example of pre-existing soil being removed is through the activity of glaciers. The massive weight of the glacier scours the landscape down to the bedrock as the glacier moves. This removes any original soil and leaves exposed rock once the glacier melts and retreats.

Figure 24. During primary succession in lava on Maui, Hawaii, succulent plants are the pioneer species. (credit: Forest and Kim Starr)

In both cases, the ecosystem starts with bare rock that is devoid of life. New soil is slowly formed as weathering and other natural forces break down the rock and lead to the establishment of hearty organisms, such as lichens and some plants, which are collectively known as **pioneer species** (Figure 24) because they are the first to appear. These species help to further break down the mineral-rich rock into soil where other, less hardy but more competitive species, such as grasses, shrubs, and trees, will grow and eventually replace the pioneer species. Over time the area will reach an equilibrium state, with a set of organisms quite different from the pioneer species.

Secondary succession with Example

A classic example of secondary succession occurs in forests cleared by wildfire, or by clearcut logging (Figure 25). Wildfires will burn most vegetation, and unless the animals can flee the area, they are killed. Their nutrients, however, are returned to the ground in the form of ash. Thus, although the community has been dramatically altered, there is a soil ecosystem present that provides a foundation for rapid recolonization.

Before the fire, the vegetation was dominated by tall trees with access to the major plant energy resource: sunlight. Their height gave them access to sunlight while also shading the ground and other low-lying species. After the fire, though, these trees are no longer dominant. Thus, the first plants to grow back are usually annual plants followed within a few years by quickly growing and spreading grasses and other pioneer species. Due, at least in part, to changes in the environment brought on by the growth of grasses and forbs, over many years, shrubs emerge along with small trees. These organisms are called intermediate species. Eventually, over 150 years or more, the forest will reach its equilibrium point and resemble the community before the fire. This equilibrium state is referred to as the **climax community**, which will remain until the next disturbance. The climax community is typically characteristic of a given climate and geology. Although the community in equilibrium looks the same once it is attained, the equilibrium is a dynamic one with constant changes in abundance and sometimes species identities.

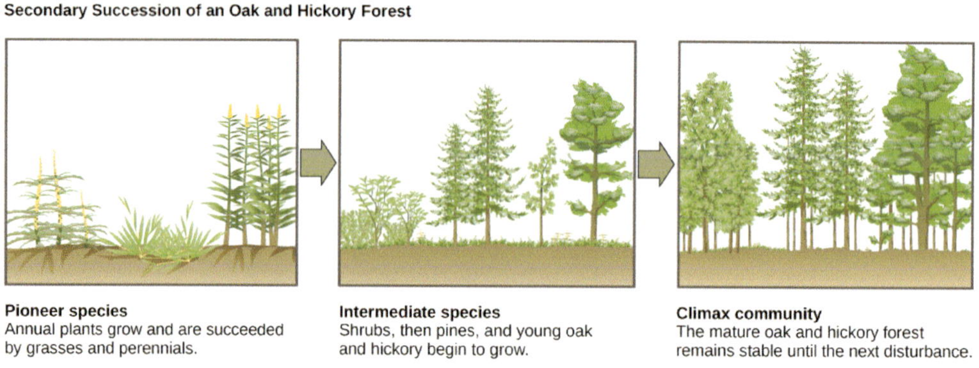

Figure 25. Secondary succession is seen in an oak and hickory forest after a forest fire. A sequence of the community present at three successive times at the same location is depicted.

AT A GLANCE

Populations are individuals of a species that live in a particular habitat. Ecologists measure characteristics of populations: size, density, and distribution pattern. Life tables are useful to calculate life expectancies of individual population members. Survivorship curves show the number of individuals surviving at each age interval plotted versus time. Populations with unlimited resources grow exponentially—with an accelerating growth rate. When resources become limiting, populations follow a logistic growth curve in which population size will level off at the carrying capacity. Humans have increased their carrying capacity through technology, urbanization, and harnessing the energy of fossil fuels. Unchecked human population growth could have dire long-term effects on human welfare and Earth's ecosystems. Communities include all the different species living in a given area. The variety of these species is referred to as biodiversity. Species may form symbiotic relationships such as commensalism, mutualism, or parasitism. Community structure is described by its foundation and keystone species. Communities respond to environmental disturbances by succession: the predictable appearance of different types of plant species, until a stable community structure is established.

CHAPTER-5

CONSERVATION OF BIODIVERSITY

5.1 BIODIVERSITY: AN INTRODUCTION

Earth is home to an impressive array of life forms. From single-celled organisms to creatures made of many trillions of cells, life has taken on many wonderful shapes and evolved countless strategies for survival. Recall that cell theory dictates that all living things are made of one or more cells. Some organisms are made of just a single cell, and are thus referred to as **unicellular**. Organisms containing more than one cell are said to be **multicellular.** Despite the wide range of organisms, there exists only two fundamental cell plans: prokaryotic and eukaryotic. The main difference these two cell plans is that eukaryotic cells have internal, membrane-bound structures called organelles (see chp 2.3). Thus, if you were to microscopically analyze the cells of any organism on Earth, you would find either prokaryotic or eukaryotic cells depending on the type of organism.

Biologists name, group, and classify organisms based on similarities in genetics and morphology. This branch of biological science is known as **taxonomy**. Taxonomists group organisms into categories that range from very broad to very specific (Figure 1). The broadest category is called **domain** and the most specific is **species** (notice the similarities between the words *specific* and *species*). Currently, taxonomists recognize three domains: Bacteria, Archaea, and Eukarya. All life forms are classified within these three domains.

Domain Bacteria

Domain Bacteria includes prokaryotic, unicellular organisms (Figure 2). They are incredibly abundant and found in nearly every imaginable type of habitat, including your body. While many people view bacteria only as disease-causing organisms, most species are actually either benign or beneficial to humans. While it is true that some bacteria may cause disease in people, this is more the exception than the rule.

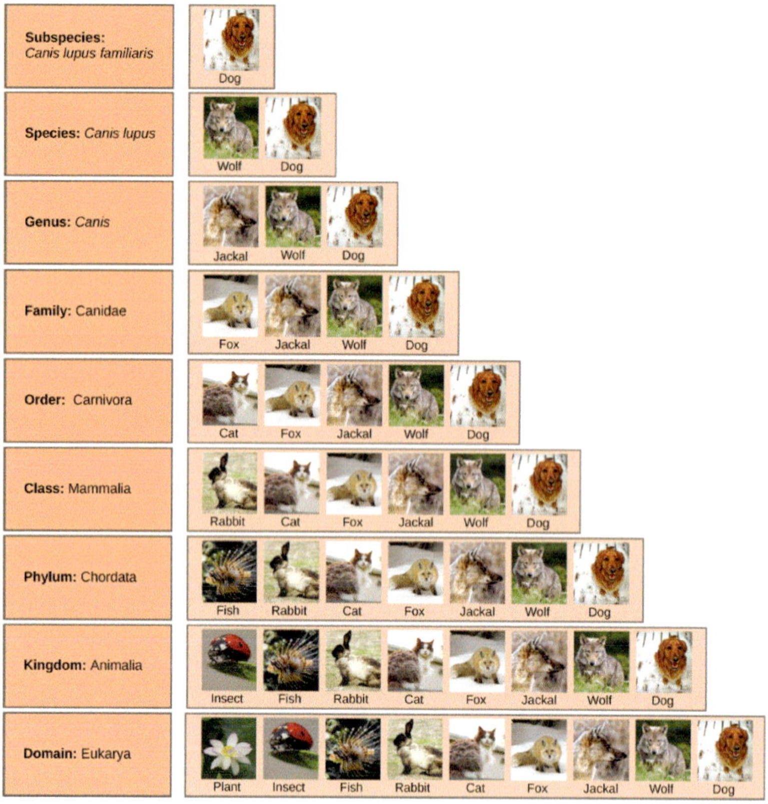

Figure 1. This illustration shows the taxonomic groups, in sequence, with examples.

Bacteria are well-known for their metabolic diversity. Metabolism is a general term describing the complex biochemistry that occurs inside of cells. Many species of bacteria are autotrophs, meaning they can create their own food source without having to eat other organisms. Most autotrophic bacteria do this by using photosynthesis, a process that converts light energy into chemical energy that can be utilized by cells. A well-known and ecologically-important group of photosynthetic bacteria is cyanobacteria. These are sometimes referred to a blue-green algae, but this name is not appropriate because, as you will see shortly, algae are organisms that belong to domain Eukarya. Cyanobacteria play important roles in food webs of aquatic systems, such as lakes.

Other species of bacteria are heterotrophs, meaning that they need to acquire their food by eating other organisms. This classification includes the bacteria that cause disease in humans (during an infection, the bacteria is eating you). However, most heterotrophic bacteria are harmless to humans. In fact, you have hundreds of species of bacteria living on your skin and in your large intestine that do you no harm. Beyond your body, heterotrophic bacteria play vital roles in ecosystems, especially soil-dwelling bacteria that decompose living matter and make nutrients available to plants.

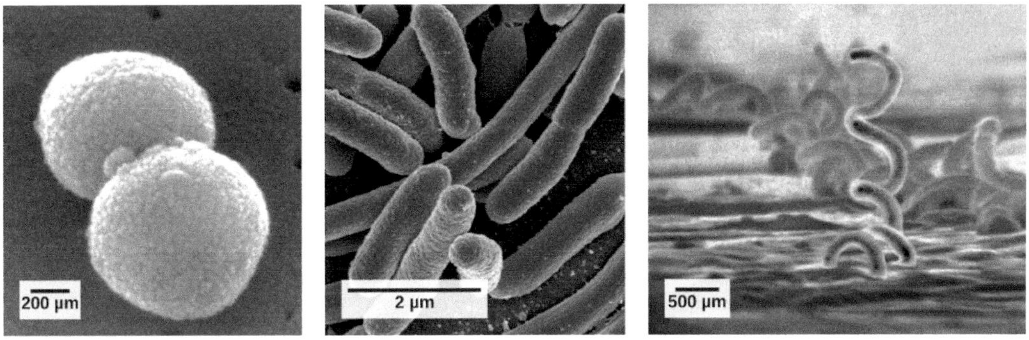

Figure 2. Many prokaryotes fall into three basic categories based on their shape: (a) cocci, or spherical; (b) bacilli, or rod-shaped; and (c) spirilla, or spiral-shaped. (credit a: modification of work by Janice Haney Carr, Dr. Richard Facklam, CDC; credit c: modification of work by Dr. David Cox, CDC; scale-bar data from Matt Russell). This figure by OpenStax is licensed under CC BY 4.0

Domain Archaea

Like bacteria, organisms in domain Archaea are prokaryotic and unicellular. Superficially, they look a lot like bacteria, and many biologists confused them as bacteria until a few decades ago. But hiding in their genes is a story that modern DNA analysis has recently revealed: archaeans are so different genetically that they belong in their own domain.

Many archaean species are found in some of the most inhospitable environments, areas of immense pressure (bottom of the ocean), salinity (such as the Great Salt Lake), or heat (geothermal springs). Organisms that can tolerate and even thrive in such conditions are known as extremophiles. (It should be noted that many bacteria are also extremophiles). Along with genetic evidence, the fact that a large percentage of archaeans are extremophiles suggests that they may be descendants of some of the most ancient lifeforms on Earth; life that originated on a young planet that was inhospitable by today's standards.

For whatever reason, archaeans are not as abundant in and on the human body as bacteria, and they cause substantially fewer diseases. Research on archaeans continues to shed light on this interesting and somewhat mysterious domain.

Domain Eukarya

This domain is most familiar to use because it includes humans and other animals, along with plants, fungi, and a lesser-known group, the protists. Unlike the other domains, Domain Eukarya contains multicellular organisms, in addition to unicellular species. The domain is characterized by the presence of eukaryotic cells.

For this domain, you will be introduced to several of its kingdoms. Kingdom is the taxonomic grouping immediately below domain (see Figure 1).

Kingdom Animalia is comprised of multicellular, heterotrophic organisms. This kingdom includes humans and other primates, insects, fish, reptiles, and many other types of animals. Kingdom Plantae includes multicellular, autotrophic organisms. Except for a few species that are parasites, plants use photosynthesis to meet their energy demands.

Figure 3. The (a) familiar mushroom is only one type of fungus. The brightly colored fruiting bodies of this (b) coral fungus are displayed. This (c) electron micrograph shows the spore-bearing structures of Aspergillus, a type of toxic fungi found mostly in soil and plants. (credit a: modification of work by Chris Wee; credit b: modification of work by Cory Zanker; credit c: modification of work by Janice Haney Carr, Robert Simmons, CDC; scale-bar data from Matt Russell). This work by OpenStax is licensed under CC BY 4.0.

Kingdom Fungi includes multicellular and unicellular, heterotrophic fungi. Fungi are commonly mistaken for plants because some species of fungi grow in the ground. Fungi are fundamentally different from plants in that they do not perform photosynthesis and instead feed on the living matter of others. Another misconception is that all fungi are mushrooms. A mushroom is a temporary reproductive structure used by some fungal species, but not all. Some fungi take the form of molds and mildews, which are commonly seen on rotting food. Lastly, yeast are unicellular fungi. Many species of yeast are important to humans, especially baker's and brewer's yeast. Through their metabolism, these yeast produce $CO2$ gas and alcohol. The former makes bread rise and the latter is the source for all alcoholic beverages.

Protists refer to a highly disparate group that was formerly its own kingdom until recent genetic analysis indicated that it should be split in to many kingdoms (Figure 4). As a group, protists are very diverse and include unicellular, multicellular, heterotrophic, and autotrophic organisms. The term 'protist' was used as a catchall for any eukaryote that was neither animal, plant, or fungus. Examples of protists include macroalgae such as kelps and seaweeds, microalgae such as diatoms and dinoflagellates, and important disease-causing microbes such as Plasmodium, the parasite that causes malaria. Sadly, malaria kills hundreds of thousands of people every year.

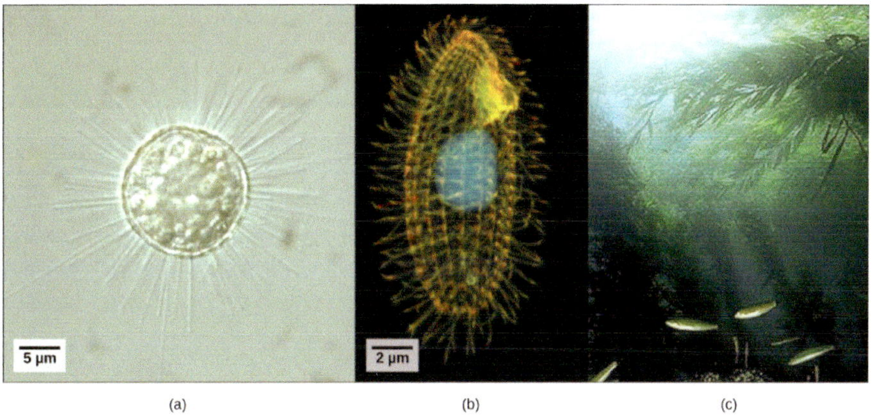

Figure 4. Protists range from the microscopic, single-celled (a) Acanthocystis turfacea and the (b) ciliate Tetrahymena thermophila to the enormous, multicellular (c) kelps (Chromalveolata) that extend for hundreds of feet in underwater "forests." (credit a: modification of work by Yuiuji Tsukii; credit b: modification of work by Richard Robinson, Public Library of Science; credit c: modification of work by Kip Evans, NOAA; scale-bar data from Matt Russell). This work by OpenStax is licensed under CC BY 4.0

With this cursory and fundamental understanding of biological diversity, you are now better equipped to study the role of biodiversity in the biosphere and in human economics, health, and culture. Each life form, even the smallest microbe, is a fascinating and and complex living machine. This complexity means we will likely never fully understand each organism and the myriad ways they interact with each other, with us, and with their environment. Thus, it is wise to value biodiversity and take measures to conserve it.

5.2 EVOLUTION AND ORIGIN OF BIODIVERSITY

Biological Process: Responsible for Biodiversity

All species —from the bacteria on our skin to the birds outside—evolved at some point from a different species. Although it may seem that living things today stay much the same from generation to generation, that is not the case because evolution is ongoing. Evolution is the process through which the characteristics of a species change over time, which can ultimately cause new species to arise.

The theory of evolution is the unifying theory of biology, meaning it is the framework within which biologists ask questions about the living world. Its power is that it provides direction for predictions about living things that are borne out in experiment after experiment. The Ukrainian-born American geneticist Theodosius Dobzhansky famously wrote that "nothing makes sense in biology except in the light of evolution." He meant that the principle that all life has evolved and diversified from a common ancestor is the foundation from which we understand all other questions in biology.

Figure 5. The diversity of life on Earth is the result of evolution, a continuous process that is still occurring. (credit "wolf": modification of work by Gary Kramer, USFWS; credit "coral": modification of work by William Harrigan, NOAA; credit "river": modification of work by Vojtěch Dostál; credit "protozoa": modification of work by Sharon Franklin, Stephen Ausmus, USDA ARS; credit "fish" modification of work by Christian Mehlführer; credit "mushroom", "bee": modification of work by Cory Zanker; credit "tree": modification of work by Joseph Kranak)

Discovering How Populations Change: The Actual Mechanism

The theory of evolution by natural selection describes a mechanism for how species can change over time. That species change was suggested and debated well before Darwin. The view that species were unchanging was grounded in the writings of Plato, yet there were also ancient Greeks that expressed evolutionary ideas.

In the eighteenth century, ideas about the evolution of animals were reintroduced by the various naturalists. At the same time, James Hutton, the Scottish naturalist, proposed that geological change occurred gradually by the accumulation of small changes from processes (over long periods of time) just like those happening today. This contrasted with the predominant view that the geology of the planet was a consequence of catastrophic events occurring during a relatively brief past. Hutton's view was later popularized by the geologist Charles

Lyell in the nineteenth century. Lyell became a friend to Darwin and his ideas were very influential on Darwin's thinking. Lyell argued that the greater age of Earth gave more time for gradual change in species, and the process provided an analogy for gradual change in species.

Charles Darwin and Natural Selection

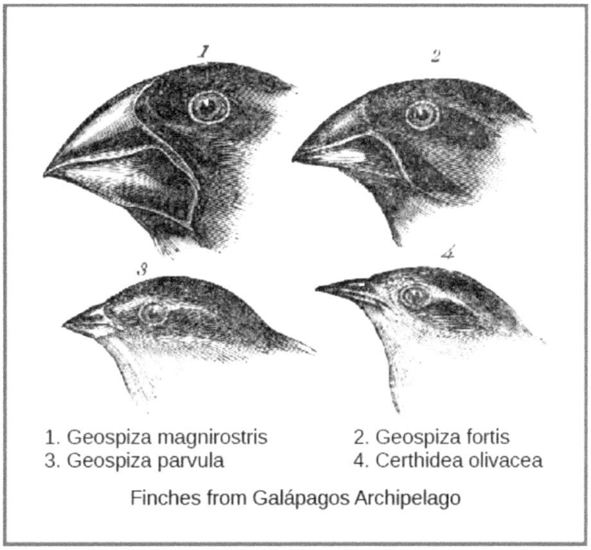

Figure 6. Darwin observed that beak shape varies among finch species. He postulated that the beak of an ancestral species had adapted over time to equip the finches to acquire different food sources. This illustration shows the beak shapes for four species of ground finch: 1. Geospiza magnirostris (the large ground finch), 2. G. fortis (the medium ground finch), 3. G. parvula (the small tree finch), and 4. Certhidea olivacea (the green-warbler finch).

Natural selection as a mechanism for evolution was independently conceived of and described by two naturalists, Charles Darwin and Alfred Russell Wallace, in the mid-nineteenth century. Importantly, each spent years exploring the natural world on expeditions to the tropics. From 1831 to 1836, Darwin traveled around the world on H.M.S. Beagle, visiting South America, Australia, and the southern tip of Africa. Wallace traveled to Brazil to collect insects in the Amazon rainforest from 1848 to 1852 and to the Malay Archipelago from 1854 to 1862. Darwin's journey, like Wallace's later journeys in the Malay Archipelago, included stops at several island chains, the last being the Galápagos Islands (west of Ecuador). On these islands, Darwin observed species of organisms on different islands that were clearly similar, yet had distinct differences. For example, the ground finches inhabiting the Galápagos Islands comprised several species that each had a unique beak shape (Figure 2). He observed both that these finches closely resembled another finch species on the mainland of South America and that the group of species in the Galápagos formed a graded series of beak sizes and shapes, with very small differences between the most similar. Darwin imagined that the island species might be all species modified from one original mainland species. In 1860, he wrote, "Seeing this gradation and diversity of structure in one small, intimately related group of birds, one might really fancy that from an original paucity of birds in this archipelago, one species had been taken and modified for different ends."

Wallace and Darwin both observed similar patterns in other organisms and independently conceived a mechanism to explain how and why such changes could take place. Darwin called this mechanism natural selection. Natural selection, Darwin argued, was an inevitable outcome of three principles that operated in nature. First, there exists variation in traits among individuals within a population, and these traits are inherited, or passed from parent to offspring. Second, more offspring are produced than are able to survive; in other words, resources for survival and reproduction are limited. And lastly, there is a competition for those resources in each generation. Out of these three principles, Darwin and Wallace reasoned that offspring with inherited characteristics that allow them to best compete for limited resources will survive and have more offspring than those individuals with variations that are less able to compete. Because characteristics are inherited, these traits will be better represented in the next generation. This will lead to change in populations over generations in a process that Darwin called "descent with modification." In sum, we can define natural selection as a process the causes beneficial traits to become more common in a population over time, causing the population to evolve.

Papers by Darwin and Wallace (Figure 3) presenting the idea of natural selection were read together in 1858 before the Linnaean Society in London. The following year Darwin's book, *On the Origin of Species,* was published, which outlined in considerable detail his arguments for evolution by natural selection.

Natural selection can only take place if there is **variation**, or differences, among individuals in a population. Importantly, these differences must have some genetic basis, otherwise natural selection would not lead to change in the next generation because there would be no way to transmit those traits from one generation to the next.

Genetic diversity in a population comes from two main sources: mutation and sexual reproduction. Mutation, a permanent change in DNA sequence is the ultimate source of new genetic variation in any population. An individual that has a mutated gene might have a different trait than other individuals in the population. Without variation in traits, nature would not be able to select the traits that are best adapted for the organisms' environment at that particular time.

(a) (b)

Figure 7. (a) Charles Darwin and (b) Alfred Wallace wrote scientific papers on natural selection that were presented together before the Linnean Society in 1858.

Evolutionary change in action

The development of antibiotic resistant bacteria is an example of evolution through natural selection and it has been directly observed by scientists. How does this happen? Imagine a person that has a bacterial infection: their body is being attacked by billions of bacteria. Because there is genetic variation in populations, some individual bacteria may already possess traits that allow them to tolerate antibiotic drugs. When the infected person is prescribed antibiotics, the drug attacks and kills the entire population, except for those bacteria that can resist the drug. These bacteria survive because they had a trait that was beneficial and thus nature selected for it. The surviving population will all be resistant to the drug and continue to reproduce, multiple, and pass down that beneficial trait to all offspring. The population has now evolved because all individuals have the antibiotic-resistant trait, whereas before it was rare. It is important to realize that evolution occurs at the population level and is reliant upon genetic variation that was already present. Without that variation, there is nothing for nature to select for. The rise and spread of antibiotic resistant bacteria is an emerging environmental issue and will be discussed in a later chapter.

5.3 SIGNIFICANCE OF BIODIVERSITY

The Biodiversity Crisis

Biologists estimate that species extinctions are currently 500–1000 times the normal, or background, rate seen previously in Earth's history. The current high rates will cause a precipitous decline in the biodiversity of the planet in the next century or two. The loss of biodiversity will include many species we know today.

Although it is sometimes difficult to predict which species will become extinct, many are listed as endangered (at great risk of extinction). However, many extinctions will affect species that biologist have not yet discovered. Most of these "invisible" species that will become extinct currently live in tropical rainforests like those of the Amazon basin. These rainforests are the most diverse ecosystems on the planet and are being destroyed rapidly by deforestation. Between 1970 and 2011, almost 20 percent of the Amazon rainforest was lost.

Figure 8. This tropical lowland rainforest in Madagascar is an example of a high biodiversity habitat. This particular location is protected within a national forest, yet only 10 percent of the original coastal lowland forest remains, and research suggests half the original biodiversity has been lost. (credit: Frank Vassen)

Biodiversity is a broad term for biological variety, and it can be measured at a number of organizational levels. Traditionally, ecologists have measured biodiversity by taking into account both the number of species and the number of individuals of each species (known as **relative abundance**). However, biologists are using different measures of biodiversity, including genetic diversity, to help focus efforts to preserve the biologically and technologically important elements of biodiversity.

Biodiversity loss refers to the reduction of biodiversity due to displacement or extinction of species. The loss of a particular individual species may seem unimportant to some, especially if it is not a charismatic species like the Bengal tiger or the bottlenose dolphin. However, the current accelerated extinction rate means the loss of tens of thousands of species within our lifetimes. Much of this loss is occurring in tropical rainforests like the one pictured in Figure 1, which are very high in biodiversity but are being cleared for timber and agriculture. This is likely to have dramatic effects on human welfare through the collapse of ecosystems.

Biologists recognize that human populations are embedded in ecosystems and are dependent on them, just as is every other species on the planet. Agriculture began after early hunter-gatherer societies first settled in one place and heavily modified their immediate environment. This cultural transition has made it difficult for humans to recognize their dependence on living things other than crops and domesticated animals on the planet. Today our technology smooths out the harshness of existence and allows many of us to live longer, more comfortable lives, but ultimately the human species cannot exist without its surrounding ecosystems. Our ecosystems provide us with food, medicine, clean air and water, recreation, and spiritual and aesthetical inspiration.

Different Types of Biodiversity

A common meaning of biodiversity is simply the number of species in a location or on Earth; for example, the American Ornithologists' Union lists 2078 species of birds in North and Central America. This is one measure of the bird biodiversity on the continent. More sophisticated measures of diversity take into account the relative abundances of species. For example, a forest with 10 equally common species of trees is more diverse than a forest that has 10 species of trees wherein just one of those species makes up 95 percent of the trees. Biologists have also identified alternate measures of biodiversity, some of which are important in planning how to preserve biodiversity.

Genetic diversity is one alternate concept of biodiversity. **Genetic diversity** is the raw material for evolutionary adaptation in a species and is represented by the variety of genes present within a population. A species' potential to adapt to changing environments or new diseases depends on this genetic diversity.

It is also useful to define **ecosystem diversity**: the number of different ecosystems on Earth or in a geographical area. The loss of an ecosystem means the loss of the interactions between species and the loss of biological productivity that an ecosystem is able to create. An example of a largely extinct ecosystem in North America is the prairie ecosystem (Figure 2). Prairies once spanned central North America from the boreal forest in northern Canada down into Mexico. They are now all but gone, replaced by crop fields, pasture lands, and suburban sprawl. Many of the species survive, but the hugely productive ecosystem that was responsible for creating our most productive agricultural soils is now gone. As a consequence, their soils are now being depleted unless they are maintained artificially at great expense. The decline in soil productivity occurs because the interactions in the original ecosystem have been lost.

Figure 9. The variety of ecosystems on Earth—from coral reef to prairie—enables a great diversity of species to exist. (credit "coral reef": modification of work by Jim Maragos, USFWS; credit: "prairie": modification of work by Jim Minnerath, USFWS)

Current Trends in Species Diversity

Despite considerable effort, knowledge of the species that inhabit the planet is limited. A recent estimate suggests that only 13% of eukaryotic species have been named (Table 1). Estimates of numbers of prokaryotic species are largely guesses, but biologists agree that science has only just begun to catalog their diversity. . Given that Earth is losing species at an accelerating pace, science knows little about what is being lost.

Table 1. This table shows the estimated number of species by taxonomic group—including both described (named and studied) and predicted (yet to be named) species.

Estimated Numbers of Described and Predicted species						
	Source: Mora et al 2011		Source: Chapman 2009		Source: Groombridge and Jenkins 2002	
	Described	Predicted	Described	Predicted	Described	Predicted
Animals	1,124,516	9,920,000	1,424,153	6,836,330	1,225,500	10,820,000
Photosynthetic Protists	17,892	34,900	25,044	200,500	—	—
Fungi	44,368	616,320	98,998	1,500,000	72,000	1,500,000
Plants	224,244	314,600	310,129	390,800	270,000	320,000
Non-Photosynthetic Protists	16,236	72,800	28,871	1,000,000	80,000	600,000
Prokaryotes	—	—	10,307	1,000,000	10,175	—
Total	1,438,769	10,960,000	1,897,502	10,897,630	1,657,675	13,240,000

There are various initiatives to catalog described species in accessible and more organized ways, and the internet is facilitating that effort. Nevertheless, at the current rate of species description, which according to the State of Observed

Species1 reports is 17,000–20,000 new species a year, it would take close to 500 years to describe all of the species currently in existence. The task, however, is becoming increasingly impossible over time as extinction removes species from Earth faster than they can be described.

Naming and counting species may seem an unimportant pursuit given the other needs of humanity, but it is not simply an accounting. Describing species is a complex process by which biologists determine an organism's unique characteristics and whether or not that organism belongs to any other described species. It allows biologists to find and recognize the species after the initial discovery to follow up on questions about its biology. That subsequent research will produce the discoveries that make the species valuable to humans and to our ecosystems. Without a name and description, a species cannot be studied in depth and in a coordinated way by multiple scientists.

Different Types of Biodiversity Pattern in Species

Biodiversity is not evenly distributed on the planet. Lake Victoria contained almost 500 species of cichlids (just one family of fishes that are present in the lake) before the introduction of an exotic species in the 1980s and 1990s caused a mass extinction. All of these species were found only in Lake Victoria, which is to say they were endemic. **Endemic species** are found in only one location. For example, the blue jay is endemic to North America, while the Barton Springs salamander is endemic to the mouth of one spring in Austin, Texas. Endemic species with highly restricted distributions, like the Barton Springs salamander, are particularly vulnerable to extinction.

Lake Huron contains about 79 species of fish, all of which are found in many other lakes in North America. What accounts for the difference in diversity between Lake Victoria and Lake Huron? Lake Victoria is a tropical lake, while Lake Huron is a temperate lake. Lake Huron in its present form is only about 7,000 years old, while Lake Victoria in its present form is about 15,000 years old. These two factors, latitude and age, are two of several hypotheses biogeographers have suggested to explain biodiversity patterns on Earth.

Biogeography is the study of the distribution of the world's species both in the past and in the present. The work of biogeographers is critical to understanding our physical environment, how the environment affects species, and how changes in environment impact the distribution of a species.

There are three main fields of study under the heading of biogeography: ecological biogeography, historical biogeography (called paleobiogeography), and conservation biogeography. Ecological biogeography studies the current factors affecting the distribution of plants and animals. Historical biogeography, as the name implies, studies the past distribution of species. Conservation biogeography, on the other hand, is focused on the protection and restoration of species based upon the known historical and current ecological information.

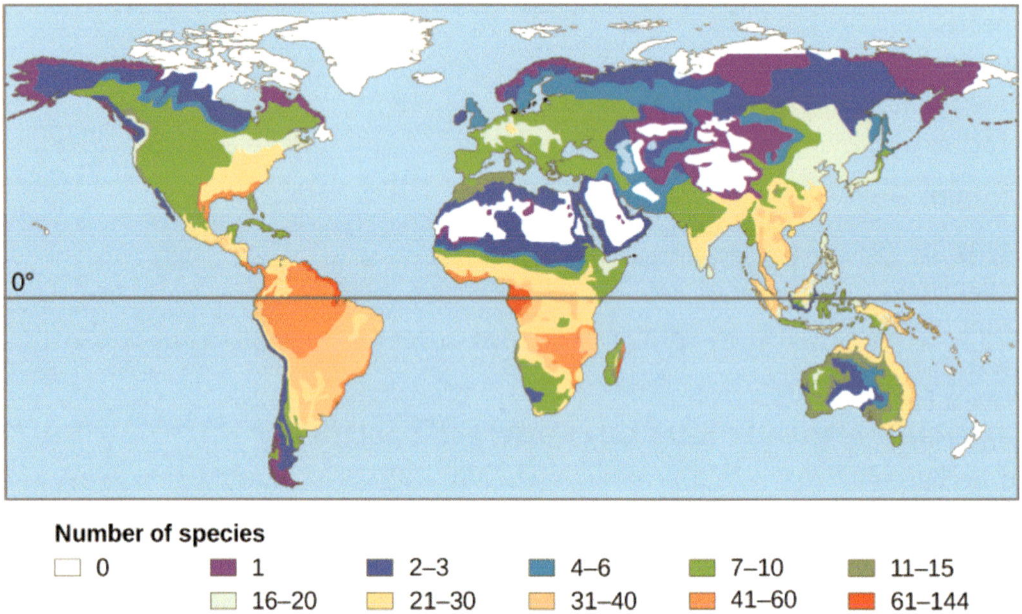

Figure 10. This map illustrates the number of amphibian species across the globe and shows the trend toward higher biodiversity at lower latitudes. A similar pattern is observed for most taxonomic groups.

One of the oldest observed patterns in ecology is that biodiversity typically increases as latitude declines. In other words, biodiversity increases closer to the equator (Figure 3).

It is not yet clear why biodiversity increases closer to the equator, but hypotheses include the greater age of the ecosystems in the tropics versus temperate regions, which were largely devoid of life or drastically impoverished during the last ice age. The greater age provides more time for **speciation,** the evolutionary process of creating new species. Another possible explanation is the greater energy the tropics receive from the sun. But scientists have not been able to explain how greater energy input could translate into more species. The complexity of tropical ecosystems may promote speciation by increasing the habitat complexity, thus providing more ecological niches. Lastly, the tropics have been perceived as being more stable than temperate regions, which have a pronounced climate and day-length seasonality. The stability of tropical ecosystems might promote speciation. Regardless of the mechanisms, it is certainly true that biodiversity is greatest in the tropics. There are also high numbers of endemic species.

Importance of Biodiversity in Human Life

Loss of biodiversity may have reverberating consequences on ecosystems because of the complex interrelations among species. For example, the extinction of one species may cause the extinction of another. Biodiversity is important to the survival and welfare of human populations because it has impacts on our health and our ability to feed ourselves through agriculture and harvesting populations of wild animals.

Human Health

Many medications are derived from natural chemicals made by a diverse group of organisms. For example, many plants produce compounds meant to protect the plant from insects and other animals that eat them. Some of these compounds also work as human medicines. Contemporary societies that live close to the land often have a broad knowledge of the medicinal uses of plants growing in their area. For centuries in Europe, older knowledge about the medical uses of plants was compiled in herbals—books that identified the plants and their uses. Humans are not the only animals to use plants for medicinal reasons. The other great apes, orangutans, chimpanzees, bonobos, and gorillas have all been observed self-medicating with plants.

Modern pharmaceutical science also recognizes the importance of these plant compounds. Examples of significant medicines derived from plant compounds include aspirin, codeine, digoxin, atropine, and vincristine (Figure 4). Many medications were once derived from plant extracts but are now synthesized. It is estimated that, at one time, 25 percent of modern drugs contained at least one plant extract. That number has probably decreased to about 10 percent as natural plant ingredients are replaced by synthetic versions of the plant compounds. Antibiotics, which are responsible for extraordinary improvements in health and lifespans in developed countries, are compounds largely derived from fungi and bacteria.

Figure 11. Catharanthus roseus, the Madagascar periwinkle, has various medicinal properties. Among other uses, it is a source of vincristine, a drug used in the treatment of lymphomas. (credit: Forest and Kim Starr)

In recent years, animal venoms and poisons have excited intense research for their medicinal potential. By 2007, the FDA had approved five drugs based on animal toxins to treat diseases such as hypertension, chronic pain, and diabetes. Another five drugs are undergoing clinical trials and at least six drugs are being used in other countries. Other toxins under investigation come from mammals, snakes, lizards, various amphibians, fish, snails, octopuses, and scorpions.

Aside from representing billions of dollars in profits, these medications improve people's lives. Pharmaceutical companies are actively looking for new natural compounds that can function as medicines. It is estimated that one third of pharmaceutical research and development is spent on natural compounds and that about 35 percent of new drugs brought to market between 1981 and 2002 were from natural compounds.

Finally, it has been argued that humans benefit psychologically from living in a biodiverse world. The chief proponent of this idea is famed entomologist E. O. Wilson. He argues that human evolutionary history has adapted us to living in a natural environment and that built environments generate stresses that affect human health and well-being. There is considerable research into the psychologically regenerative benefits of natural landscapes that suggest the hypothesis may hold some truth.

Agricultural

Since the beginning of human agriculture more than 10,000 years ago, human groups have been breeding and selecting crop varieties. This crop diversity matched the cultural diversity of highly subdivided populations of humans. For example, potatoes were domesticated beginning around 7,000 years ago in the central Andes of Peru and Bolivia. The people in this region traditionally lived in relatively isolated settlements separated by mountains. The potatoes grown in that region belong to seven species and the number of varieties likely is in the thousands. Each variety has been bred to thrive at particular elevations and soil and climate conditions. The diversity is driven by the diverse demands of the dramatic elevation changes, the limited movement of people, and the demands created by crop rotation for different varieties that will do well in different fields.

Potatoes are only one example of agricultural diversity. Every plant, animal, and fungus that has been cultivated by humans has been bred from original wild ancestor species into diverse varieties arising from the demands for food value, adaptation to growing conditions, and resistance to pests. The potato demonstrates a well-known example of the risks of low crop diversity: during the tragic Irish potato famine (1845–1852 AD), the single potato variety grown in Ireland became susceptible to a potato blight—wiping out the crop. The loss of the crop led to famine, death, and mass emigration. Resistance to disease is a chief benefit to maintaining crop biodiversity and lack of diversity in contemporary crop species carries similar risks. Seed companies, which are the source of most crop varieties in developed countries, must continually breed new varieties to keep up with evolving pest organisms. These same seed companies, however, have participated in the decline of the number of varieties available as they focus on selling fewer varieties in more areas of the world replacing traditional local varieties.

The ability to create new crop varieties relies on the diversity of varieties available and the availability of wild forms related to the crop plant. These wild forms are often the source of new gene variants that can be bred with existing varieties to create varieties with new attributes. Loss of wild species related to a crop will mean the loss of potential in crop improvement. Maintaining the genetic diversity of wild species related to domesticated species ensures our continued supply of food.

Since the 1920s, government agriculture departments have maintained seed banks of crop varieties as a way to maintain crop diversity. This system has flaws because over time seed varieties are lost through accidents and there is no way to replace them. In 2008, the Svalbard Global seed Vault, located on Spitsbergen island, Norway, (Figure) began storing seeds from around the world as a backup system to the regional seed banks. If a regional seed bank stores varieties in Svalbard, losses can be replaced from Svalbard should something happen to the regional seeds. The Svalbard seed vault is deep into the rock of the arctic island. Conditions within the vault are maintained at ideal temperature and humidity for seed survival, but the deep underground location of the vault in the arctic means that failure of the vault's systems will not compromise the climatic conditions inside the vault.

Figure 12. The Svalbard Global Seed Vault is a storage facility for seeds of Earth's diverse crops. (credit: Mari Tefre, Svalbard Global Seed Vault)

Although crops are largely under our control, our ability to grow them is dependent on the biodiversity of the ecosystems in which they are grown. That biodiversity creates the conditions under which crops are able to grow through what are known as ecosystem services—valuable conditions or processes that are carried out by an ecosystem. Crops are not grown, for the most part, in built environments. They are grown in soil. Although some agricultural soils are rendered sterile using controversial pesticide treatments, most contain a huge diversity of organisms that maintain nutrient cycles—breaking down organic matter into nutrient compounds that crops need for growth. These organisms also maintain soil texture that affects water and oxygen dynamics in the soil that are necessary for plant growth. Replacing the work of these organisms in forming arable soil is not practically possible. These kinds of processes are called ecosystem services. They occur within ecosystems, such as soil ecosystems, as a result of the diverse metabolic activities of the organisms living there, but they provide benefits to human food production, drinking water availability, and breathable air.

Other key ecosystem services related to food production are plant pollination and crop pest control. It is estimated that honeybee pollination within the United States brings in $1.6 billion per year; other pollinators contribute up to $6.7 billion. Over 150 crops in the United States require pollination to produce. Many honeybee populations are managed by beekeepers who rent out their hives' services to farmers. Honeybee populations in North America have been suffering large losses caused by a syndrome known as colony collapse disorder, a new phenomenon with an unclear cause. Other pollinators include a diverse array of other bee species and various insects and birds. Loss of these species would make growing crops requiring pollination impossible, increasing dependence on other crops.

Finally, humans compete for their food with crop pests, most of which are insects. Pesticides control these competitors, but these are costly and lose their effectiveness over time as pest populations adapt. They also lead to collateral damage by killing non-pest species as well as beneficial insects like honeybees, and risking the health of agricultural workers and consumers. Moreover, these pesticides may migrate from the fields where they are applied and do damage to other ecosystems like streams, lakes, and even the ocean. Ecologists believe that the bulk of the work in removing pests is actually done by predators and parasites of those pests, but the impact has not been well studied. A review found that in 74 percent of studies that looked for an effect of landscape complexity (forests and fallow fields near to crop fields) on natural enemies of pests, the greater the complexity, the greater the effect of pest-suppressing organisms. Another experimental study found that introducing multiple enemies of pea aphids (an important alfalfa pest) increased the yield of alfalfa significantly. This study shows that a diversity of pests is more effective at control than one single pest. Loss of diversity in pest enemies will inevitably make it more difficult and costly to grow food. The world's growing human population faces significant challenges in the increasing costs and other difficulties associated with producing food.

Wild Food Sources

In addition to growing crops and raising food animals, humans obtain food resources from wild populations, primarily wild fish populations. For about one billion people, aquatic resources provide the main source of animal protein. But since 1990, production from global fisheries has declined. Despite considerable effort, few fisheries on Earth are managed sustainability.

Fishery extinctions rarely lead to complete extinction of the harvested species, but rather to a radical restructuring of the marine ecosystem in which a dominant species is so over-harvested that it becomes a minor player, ecologically. In addition to humans losing the food source, these alterations affect many other species in ways that are difficult or impossible to predict. The collapse of fisheries has dramatic and long-lasting effects on local human populations that work in the fishery. In addition, the loss of an inexpensive protein source to populations that cannot afford to replace it will increase the cost of living and limit societies in other ways. In general, the fish taken from fisheries have shifted to smaller species and the larger species are overfished. The ultimate outcome could clearly be the loss of aquatic systems as food sources.

5.4 MAJOR THREATS TO BIODIVERSITY

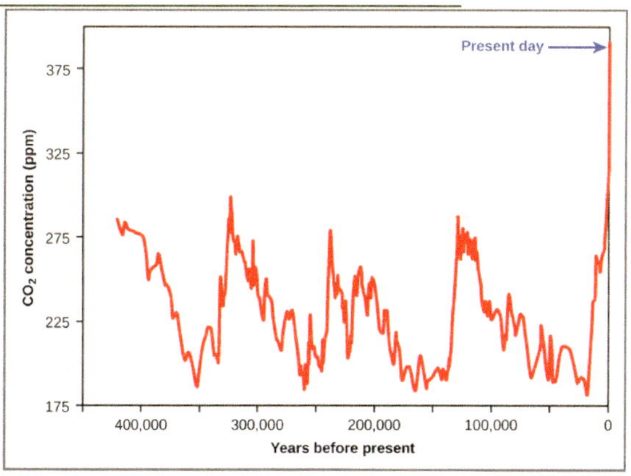

Figure 13. Atmospheric carbon dioxide levels fluctuate in a cyclical manner. However, the burning of fossil fuels in recent history has caused a dramatic increase in the levels of carbon dioxide in the Earth's atmosphere, which have now reached levels never before seen on Earth. Scientists predict that the addition of this "greenhouse gas" to the atmosphere is resulting in climate change that will significantly impact biodiversity in the coming century.

The core threat to biodiversity on the planet, and therefore a threat to human welfare, is the combination of human population growth and the resources used by that population. The human population requires resources to survive and grow, and many of those resources are being removed unsustainably from the environment. The three greatest proximate threats to biodiversity are habitat loss, overharvesting, and introduction of exotic species. The first two of these are a direct result of human population growth and resource use. The third results from increased mobility and trade. A fourth major cause of extinction, **anthropogenic** (human-caused) climate change, has not yet had a large impact, but it is predicted to become significant during this century. Global climate change is also a consequence of human population needs for energy and the use of fossil fuels to meet those needs (Figure 1). Environmental issues, such as toxic pollution, have specific targeted effects on species, but are not generally seen as threats at the magnitude of the others.

Habitat Loss

Humans rely on technology to modify their environment and make it habitable. Other species cannot do this. Elimination of their habitat—whether it is a forest, coral reef, grassland, or flowing river—will kill the individuals in the species. Remove the entire habitat and the species will become extinct, unless they are among the few species that do well in human-built environments. Human destruction of habitats (habitat generally refers to the part of the ecosystem required by a particular species) accelerated in the latter half of the twentieth century.

Figure 14. An oil palm plantation in Sabah province Borneo, Malaysia, replaces native forest habitat that a variety of species depended on to live. (credit: Lian Pin Koh)

Consider the exceptional biodiversity of Sumatra: it is home to one species of orangutan, a species of critically endangered elephant, and the Sumatran tiger, but half of Sumatra's forest is now gone. The neighboring island of Borneo, home to the other species of orangutan, has lost a similar area of forest. Forest loss continues in protected areas of Borneo. The orangutan in Borneo is listed as endangered by the International Union for Conservation of Nature (IUCN), but it is simply the most visible of thousands of species that will not survive the disappearance of the forests of Borneo. The forests are removed for timber and to plant palm oil plantations (Figure 14). Palm oil is used in many products including food products, cosmetics, and biodiesel in Europe. A 5-year estimate of global forest cover loss for the years from 2000 to 2005 was 3.1%. Much loss (2.4%) occurred in the tropics where forest loss is primarily from timber extraction. These losses certainly also represent the extinction of species unique to those areas.

Preventing Habitat Destruction with Wise Wood Choices

Most consumers do not imagine that the home improvement products they buy might be contributing to habitat loss and species extinctions. Yet the market for illegally harvested tropical timber is huge, and the wood products often find themselves in building supply stores in the United States. One estimate is that up to 10% of the imported timber in the United States, which is the world's largest consumer of wood products, is illegally logged. In 2006, this amounted to $3.6 billion in wood products. Most of the illegal products are imported from countries that act as intermediaries and are not the originators of the wood.

How is it possible to determine if a wood product, such as flooring, was harvested sustainably or even legally? The Forest Stewardship Council (FSC) certifies sustainably harvested forest products. Looking for their certification on flooring and other hardwood products is one way to ensure that the wood has not been taken illegally from a tropical forest. There are certifications other than the FSC, but these are run by timber companies, thus creating a conflict of interest. Another approach is to buy domestic wood species. While it would be great if there was a list of legal versus illegal woods, it is not that simple. Logging and forest management laws vary from country to country; what is illegal in one country may be legal in another. Where and how a product is harvested and whether the forest from which it comes is being sustainably maintained all factor into whether a wood product will be certified by the FSC. It is always a good idea to ask questions about where a wood product came from and how the supplier knows that it was harvested legally.

Habitat destruction can affect ecosystems other than forests. Rivers and streams are important ecosystems and are frequently the target of habitat modification. Damming of rivers affects flow and access to habitat. Altering a flow regime can reduce or eliminate populations that are adapted to seasonal changes in flow. For example, an estimated 91% of riverways in the United States have been modified with damming or stream bank modification. Many fish species in the United States, especially rare species or species with restricted distributions, have seen declines caused by river damming and habitat loss. Research has confirmed that species of amphibians that must carry out parts of their life cycles in both aquatic and terrestrial habitats are at greater risk of population declines and extinction because of the increased likelihood that one of their habitats or access between them will be lost. This is of particular concern because amphibians have been declining in numbers and going extinct more rapidly than many other groups for a variety of possible reasons.

Overharvesting is a Serious Threats to Many Species

Overharvesting is a serious threat to many species, but particularly to aquatic species. There are many examples of regulated fisheries (including hunting of marine mammals and harvesting of crustaceans and other species) monitored by fisheries scientists that have nevertheless collapsed. The western Atlantic cod fishery is the most spectacular recent collapse. While it was a hugely productive fishery for 400 years, the introduction of modern factory trawlers in the 1980s and the pressure on the fishery led to it becoming unsustainable. The causes of fishery collapse are both economic and political in nature.

Most fisheries are managed as a common resource, available to anyone willing to fish, even when the fishing territory lies within a country's territorial waters. Common resources are subject to an economic pressure known as the tragedy of the commons, in which fishers have little motivation to exercise restraint in harvesting a fishery when they do not own the fishery. The general outcome of harvests of resources held in common is their overexploitation. While large fisheries are regulated to attempt to avoid this pressure, it still exists in the background. This overexploitation is exacerbated when access to the fishery is open and unregulated and when technology gives fishers the ability to overfish. In a few fisheries, the biological growth of the resource is less than the potential growth of the profits made from fishing if that time and money were invested elsewhere. In these cases—whales are an example—economic forces will drive toward fishing the population to extinction.

Coral reefs are extremely diverse marine ecosystems that face peril from several processes. Reefs are home to 1/3 of the world's marine fish species—about 4000 species—despite making up only one percent of marine habitat. Most home marine aquaria house coral reef species that are wild-caught organisms—not cultured organisms. Although no marine species is known to have been driven extinct by the pet trade, there are studies showing that populations of some species have declined in response to harvesting, indicating that the harvest is not sustainable at those levels. There are also concerns about the effect of the pet trade on some terrestrial species such as turtles, amphibians, birds, plants, and even the orangutans.

Figure 15. Harvesting of pangolins for their scales and meat, and as curiosities, has led to a drastic decline in population size for this fascinating creature. This work by David Brossard is licensed under CC BY 4.0

Bush meat is the generic term used for wild animals killed for food. Hunting is practiced throughout the world, but hunting practices, particularly in equatorial Africa and parts of Asia, are believed to threaten several species with extinction. Traditionally, bush meat in Africa was hunted to feed families directly. However, recent commercialization of the practice now has bush meat available in grocery stores, which has increased harvest rates to the level of unsustainability. Additionally, human population growth has increased the need for protein foods that are not being met from agriculture. Species threatened by the bush meat trade are mostly mammals including many monkeys and the great apes living in the Congo basin.

Invasive Species

Exotic species are species that have been intentionally or unintentionally introduced by humans into an ecosystem in which they did not evolve. Human transportation of people and goods, including the intentional transport of organisms for trade, has dramatically increased the introduction of species into new ecosystems. These new introductions are sometimes at distances that are well beyond the capacity of the species to ever travel itself and outside the range of the species' natural predators.

Most exotic species introductions probably fail because of the low number of individuals introduced or poor adaptation to the ecosystem they enter. Some species, however, have characteristics that can make them especially successful in a new ecosystem. These exotic species often undergo dramatic population increases in their new habitat and reset the ecological conditions in the new environment, threatening the species that exist there. When this happens, the exotic species also becomes an **invasive species.** Invasive species can threaten other species through competition for resources, predation, or disease.

Lakes and islands are particularly vulnerable to extinction threats from introduced species. In Lake Victoria, the intentional introduction of the Nile perch was largely responsible for the extinction of about 200 species of cichlids. The accidental introduction of the brown tree snake via aircraft (Figure 4) from the Solomon Islands to Guam in 1950 has led to the extinction of three species of birds and three to five species of reptiles endemic to the island. Several other species are still threatened. The brown tree snake is adept at exploiting human transportation as a means to migrate; one was even found on an aircraft arriving in Corpus Christi, Texas. Constant vigilance on the part of airport, military, and commercial aircraft personnel is required to prevent the snake from moving from Guam to other islands in the Pacific, especially Hawaii. Islands do not make up a large area of land on the globe, but they do contain a disproportionate number of endemic species because of their isolation from mainland ancestors.

Many introductions of aquatic species, both marine and freshwater, have occurred when ships have dumped ballast water taken on at a port of origin into waters at a destination port. Water from the port of origin is pumped into tanks on a ship empty of cargo to increase stability. The water is drawn from the ocean or estuary of the port and typically contains living organisms such as plant parts, microorganisms, eggs, larvae, or aquatic animals. The water is then pumped out before the ship takes on cargo at the destination port, which may be on a different continent. The zebra mussel was introduced to the Great Lakes from Europe prior to 1988 in ballast water. The zebra mussels in the Great Lakes have created millions of dollars in clean-up costs to maintain water intakes and other facilities. The mussels have also altered the ecology of the lakes dramatically.

Figure 16. The brown tree snake, Boiga irregularis, is an exotic species that has caused numerous extinctions on the island of Guam since its accidental introduction in 1950. (credit: NPS)

They threaten native mollusk populations, but have also benefited some species, such as smallmouth bass. The mussels are filter feeders and have dramatically improved water clarity, which in turn has allowed aquatic plants to grow along shorelines, providing shelter for young fish where it did not exist before. The European green crab, *Carcinus maenas*, was introduced to San Francisco Bay in the late 1990s, likely in ship ballast water, and has spread north along the coast to Washington. The crabs have been found to dramatically reduce the abundance of native clams and crabs with resulting increases in the prey species of those native crabs.

Invading exotic species can also be disease organisms. It now appears that the global decline in amphibian species recognized in the 1990s is, in some part, caused by the fungus *Batrachochytrium dendrobatidis*, which causes the disease chytridiomycosis (Figure 5). There is evidence that the fungus is native to Africa and may have been spread throughout the world by transport of a commonly used laboratory and pet species: the African clawed frog, *Xenopus laevis*. It may well be that biologists themselves are responsible for spreading this disease worldwide. The North American bullfrog, *Rana catesbeiana*, which has also been widely introduced as a food animal but which easily escapes captivity, survives most infections of *B. dendrobatidis* and can act as a reservoir for the disease.

Figure 17. This Limosa harlequin frog (Atelopus limosus), an endangered species from Panama, died from a fungal disease called chytridiomycosis. The red lesions are symptomatic of the disease. (credit: Brian Gratwicke)

Early evidence suggests that another fungal pathogen, *Geomyces destructans*, introduced from Europe is responsible for white-nose syndrome, which infects cave-hibernating bats in eastern North America and has spread from a point of origin in western New York State (Figure 6). The disease has decimated bat populations and threatens extinction of species already listed as endangered: the Indiana bat, *Myotis sodalis*, and potentially the Virginia big-eared bat, *Corynorhinus townsendii virginianus*. How the fungus was introduced is unknown, but one logical presumption would be that recreational cavers unintentionally brought the fungus on clothes or equipment from Europe.

Climate Change

Figure 18. This little brown bat in Greeley Mine, Vermont, March 26, 2009, was found to have white-nose syndrome. (credit: modification of work by Marvin Moriarty, USFWS).

Climate change, and specifically the anthropogenic warming trend presently underway, is recognized as a major extinction threat, particularly when combined with other threats such as habitat loss. Anthropogenic warming of the planet has been observed and is due to past and continuing emission of greenhouse gases, primarily carbon dioxide and methane, into the atmosphere caused by the burning of fossil fuels and deforestation. Scientists overwhelmingly agree the present warming trend is caused by humans and some of the likely effects include dramatic and dangerous climate changes in the coming decades. Scientists predict that climate change will alter regional climates, including rainfall and snowfall patterns, making habitats less hospitable to the species living in them. The warming trend will shift colder climates toward the north and south poles, forcing species to move (if possible) with their adapted climate norms.

The shifting ranges will impose new competitive regimes on species as they find themselves in contact with other species not present in their historic range. One such unexpected species contact is between polar bears and grizzly bears. Previously, these two species had separate ranges. Now, their ranges are overlapping and there are documented cases of these two species mating and producing viable offspring. Changing climates also throw off the delicate timing adaptations that species have to seasonal food resources and breeding times. Scientists have already documented many contemporary mismatches to shifts in resource availability and timing.

Other shifts in range have been observed. For example, one study indicates that European bird species ranges have moved 91 km (56.5 mi) northward, on average. The same study suggested that the optimal shift based on warming trends was double that distance, suggesting that the populations are not moving quickly enough. Range shifts have also been observed in plants, butterflies, other insects, freshwater fishes, reptiles, amphibians, and mammals.

Climate gradients will also move up mountains, eventually crowding species higher in altitude and eliminating the habitat for those species adapted to the highest elevations. Some climates will completely disappear. The rate of warming appears to be accelerated in the arctic, which is recognized as a serious threat to polar bear populations that require sea ice to hunt seals during the winter months. Seals are a critical source of protein for polar bears. A trend to decreasing sea ice coverage has occurred since observations began in the mid-twentieth century. The rate of decline observed in recent years is far greater than previously predicted by climate models.

Figure 19. The effect of global warming can be seen in the continuing retreat of Grinnell Glacier. The mean annual temperature in Glacier National Park has increased 1.33°C since 1900. The loss of a glacier results in the loss of summer meltwaters, sharply reducing seasonal water supplies and severely affecting local ecosystems. (credit: USGS, GNP Archives)

Finally, global warming will raise ocean levels due to meltwater from glaciers and the greater volume occupied by warmer water. Shorelines will be inundated, reducing island size, which will have an effect on some species, and a number of islands will disappear entirely. Additionally, the gradual melting and subsequent refreezing of the poles, glaciers, and higher elevation mountains—a cycle that has provided freshwater to environments for centuries—will be altered. This could result in an overabundance of salt water and a shortage of fresh water.

5.5 PRESERVING BIODIVERSITY: AN EXTRA ORDINARY CHALLENGE

Preserving biodiversity is an extraordinary challenge that must be met by greater understanding of biodiversity itself, changes in human behavior and beliefs, and various preservation strategies.

Change in Biodiversity through Time

The number of species on the planet, or in any geographical area, is the result of an equilibrium of two evolutionary processes that are ongoing: speciation and extinction. When speciation rates begin to outstrip extinction rates, the number of species will increase. Likewise, the reverse is true when extinction rates begin to overtake speciation rates. Throughout the history of life on Earth, as reflected in the fossil record, these two processes have fluctuated to a greater or lesser extent, sometimes leading to dramatic changes in the number of species on the planet as reflected in the fossil record (Figure 1).

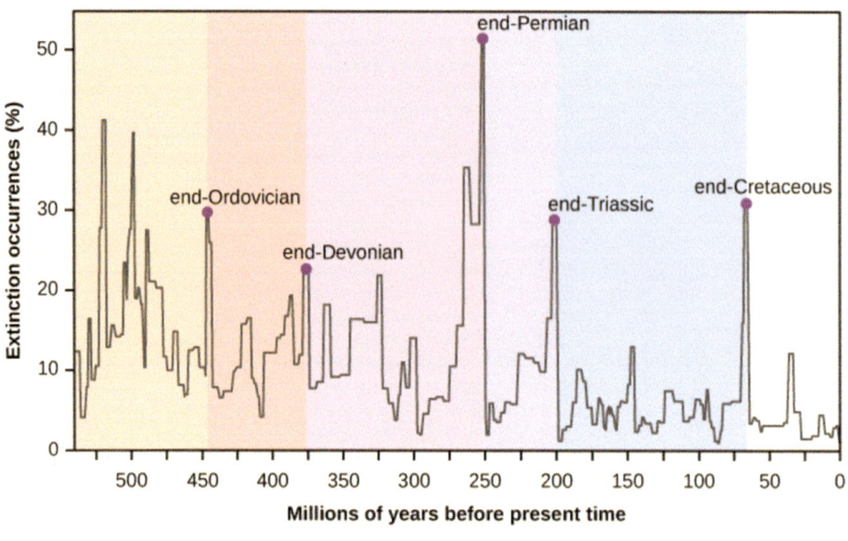

Figure 20. Extinction intensity as reflected in the fossil record has fluctuated throughout Earth's history. Sudden and dramatic losses of biodiversity, called mass extinctions, have occurred five times.

Paleontologists have identified five layers in the fossil record that appear to show sudden and dramatic losses in biodiversity. These are called **mass extinctions** and are characterized by more than half of all species disappearing from the fossil record. There are many lesser, yet still dramatic, extinction events, but the five mass extinctions have attracted the most research into their causes. An argument can be made that the five mass extinctions are only the five most extreme events in a continuous series of large extinction events throughout the fossil record (since 542 million years ago). The most recent extinction in geological time, about 65 million years ago, saw the disappearance of most dinosaurs species (except birds) and many other species. Most scientists now agree the main cause of this extinction was the impact of a large asteroid in the present-day Yucatán Peninsula and the subsequent energy release and global climate changes caused by dust ejected into the atmosphere.

Recent and Current Extinction Rates

Many biologists say that we are currently experience a sixth mass extinction and it mostly has to do with the activities of humans. There are numerous recent extinctions of individual species that are recorded in human writings. Most of these are coincident with the expansion of the European colonies since the 1500s.

One of the earlier and popularly known examples is the dodo bird. The dodo bird lived in the forests of Mauritius, an island in the Indian Ocean. The dodo bird became extinct around 1662. It was hunted for its meat by sailors and was easy prey because the dodo, which did not evolve with humans, would approach people without fear. Introduced pigs, rats, and dogs brought to the island by European ships also killed dodo young and eggs (Figure 2).

Figure 21. The dodo bird was hunted to extinction around 1662. (credit: Ed Uthman, taken in Natural History Museum, London, England)

Steller's sea cow became extinct in 1768; it was related to the manatee and probably once lived along the northwest coast of North America. Steller's sea cow was discovered by Europeans in 1741, and it was hunted for meat and oil. A total of 27 years elapsed between the sea cow's first contact with Europeans and extinction of the species. The last Steller's sea cow was killed in 1768. In another example, the last living passenger pigeon died in a zoo in Cincinnati, Ohio, in 1914. This species had once migrated in the millions but declined in numbers because of overhunting and loss of habitat through the clearing of forests for farmland.

These are only a few of the recorded extinctions in the past 500 years. The International Union for Conservation of Nature (IUCN) keeps a list of extinct and endangered species called the Red List. The list is not complete, but it describes 380 vertebrates that became extinct after 1500 AD, 86 of which were driven extinct by overhunting or overfishing.

Estimates of Current Extinction Rates

Estimates of extinction rates are hampered by the fact that most extinctions are probably happening without being observed. The extinction of a bird or mammal is often noticed by humans, especially if it has been hunted or used in some other way. But there are many organisms that are less noticeable to humans (not necessarily of less value) and many that are undescribed.

The **background extinction rate** is estimated to be about 1 per million species years (E/MSY). One "species year" is one species in existence for one year. One million species years could be one species persisting for one million years, or a million species persisting for one year. If it is the latter, then one extinction per million species years would be one of those million species becoming extinct in that year. For example, if there are 10 million species in existence, then we would expect 10 of those species to become extinct in a year. This is the background rate.

One contemporary extinction-rate estimate uses the extinctions in the written record since the year 1500. For birds alone, this method yields an estimate of 26 E/MSY, almost three times the background rate. However, this value may be underestimated for three reasons. First, many existing species would not have been described until much later in the time period and so their loss would have gone unnoticed. Second, we know the number is higher than the written record suggests because now extinct species are being described from skeletal remains that were never mentioned in written history. And third, some species are probably already extinct even though conservationists are reluctant to name them as such. Taking these factors into account raises the estimated extinction rate to nearer 100 E/MSY. The predicted rate by the end of the century is 1500 E/MSY.

A second approach to estimating present-time extinction rates is to correlate species loss with habitat loss, and it is based on measuring forest-area loss and understanding species–area relationships. The species-area relationship is the rate at which new species are seen when the area surveyed is increased (Figure 3). Likewise, if the habitat area is reduced, the number of species seen will also decline. This kind of relationship is also seen in the relationship between an island's area and the number of species present on the island: as one increases, so does the other, though not in a straight line. Estimates of extinction rates based on habitat loss and species–area relationships have suggested that with about 90 percent of habitat loss an expected 50 percent of species would become extinct. Figure 3 shows that reducing forest area from 100 km^2 to 10 km^2, a decline of 90 percent, reduces the number of species by about 50 percent. Species–area estimates have led to estimates of present-day species extinction rates of about 1000 E/MSY and higher.

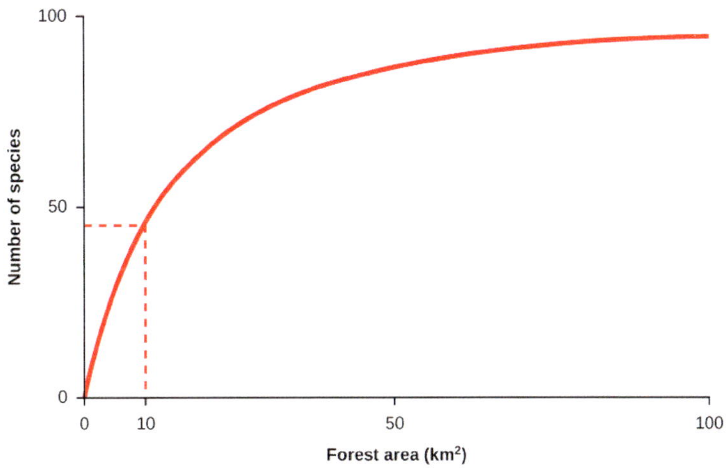

Figure 3. A typical species-area curve shows the cumulative number of species found as more and more area is sampled. The curve has also been interpreted to show the effect on species numbers of destroying habitat; a reduction in habitat of 90 percent from 100 km2 to 10 km2 reduces the number of species supported by about 50 percent.

Conservation of Biodiversity: Methods and Strategies

The threats to biodiversity have been recognized for some time. Today, the main efforts to preserve biodiversity involve legislative approaches to regulate human and corporate behavior, setting aside protected areas, and habitat restoration.

Changing Human Behavior

Legislation has been enacted to protect species throughout the world. The legislation includes international treaties as well as national and state laws. The Convention on International Trade in Endangered Species of Wild Fauna and Flora (CITES) treaty came into force in 1975. The treaty, and the national legislation that supports it, provides a legal framework for preventing "listed" species from being transported across nations' borders, thus protecting them from being caught or killed when the purpose involves international trade. The listed species that are protected by the treaty number some 33,000. The treaty is limited in its reach because it only deals with international movement of organisms or their parts. It is also limited by various countries' ability or willingness to enforce the treaty and supporting legislation. The illegal trade in organisms and their parts is probably a market in the hundreds of millions of dollars.

Within many countries there are laws that protect endangered species and that regulate hunting and fishing. In the United States, the Endangered Species Act (ESA) was enacted in 1973. When an at-risk species is listed by the Act, the U.S. Fish & Wildlife Service is required by law to develop a management plan to protect the species and bring it back to sustainable numbers. The ESA, and others like it in other countries, is a useful tool, but it suffers because it is often difficult to get a species listed or to get an effective management plan in place once a species is listed.

The Migratory Bird Treaty Act (MBTA) is an agreement between the United States and Canada that was signed into law in 1918 in response to declines in North American bird species caused by hunting. The Act now lists over 800 protected species. It makes it illegal to disturb or kill the protected species or distribute their parts (much of the hunting of birds in the past was for their feathers). Examples of protected species include northern cardinals, the red-tailed hawk, and the American black vulture.

Global warming is expected to be a major driver of biodiversity loss. Many governments are concerned about the effects of anthropogenic global warming, primarily on their economies and food resources. Because greenhouse gas emissions do not respect national boundaries, the effort to curb them is international. The international response to global warming has been mixed. The Kyoto Protocol, an international agreement that came out of the United Nations Framework Convention on Climate Change that committed countries to reducing greenhouse gas emissions by 2012, was ratified by some countries, but spurned by others. Two countries that were especially important in terms of their potential impact that did not ratify the Kyoto protocol were the United States and China. Some goals for reduction in greenhouse gasses were met and exceeded by individual countries, but, worldwide, the effort to limit greenhouse gas production is not succeeding. A renegotiated 2016 treaty, called the Paris Agreement, once again brought nations together to take meaningful action on climate change. But like before, some nations are reluctant to participate. The newly-elected President Trump has indicated that he will withdraw the United States' support of the agreement.

Conservation in Preserves

Figure 22. National parks, such as Grand Teton National Park in Wyoming, help conserve biodiversity. (credit: Don DeBold)

Establishment of wildlife and ecosystem preserves is one of the key tools in conservation efforts (Figure 23). A **preserve** is an area of land set aside with varying degrees of protection for the organisms that exist within the boundaries of the preserve. In 2003, the IUCN World Parks Congress estimated that 11.5 percent of Earth's land surface was covered by preserves of various kinds. This area is large but only represents 9 out of 14 recognized major biomes and research has shown that 12 percent of all species live outside preserves.

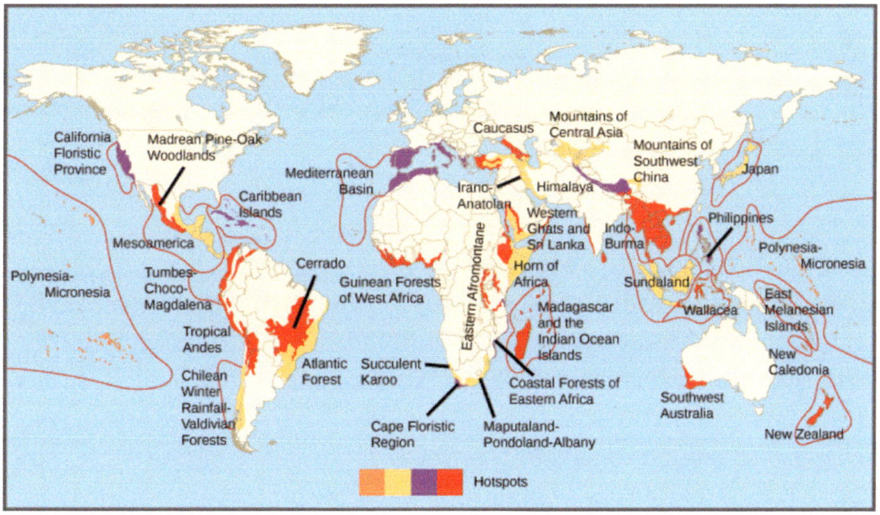

Figure 23. Conservation International has identified 34 biodiversity hotspots. Although these cover only 2.3 percent of the Earth's surface, 42 percent of the terrestrial vertebrate species and 50 percent of the world's plants are endemic to those hotspots.

A biodiversity hotspot is a conservation concept developed by Norman Myers in 1988. Hotspots are geographical areas that contain high numbers of endemic species. The purpose of the concept was to identify important locations on the planet for conservation efforts, a kind of conservation triage. By protecting hotspots, governments are able to protect a larger number of species. The original criteria for a hotspot included the presence of 1500 or more species of endemic plants and 70 percent of the area disturbed by human activity. There are now 34 biodiversity hotspots (Figure 5) that contain large numbers of endemic species, which include half of Earth's endemic plants.

There has been extensive research into optimal preserve designs for maintaining biodiversity. The fundamental principles behind much of the research have come from the seminal theoretical work of Robert H. MacArthur and Edward O. Wilson published in 1967 on island biogeography.1 This work sought to understand the factors affecting biodiversity on islands. Conservation preserves can be seen as "islands" of habitat within "an ocean" of non-habitat. In general, large preserves are better because they support more species, including species with large home ranges; they have more core area of optimal habitat for individual species; they have more niches to support more species; and they attract more species because they can be found and reached more easily. One large preserve is better than the same area of several smaller preserves because there is more core habitat unaffected by less hospitable ecosystems outside the preserve boundary. For this same reason, preserves in the shape of a square or circle will be better than a preserve with many thin "arms." If preserves must be smaller, then providing wildlife corridors (narrow strips of protected land) between two preserves is important so that species and their genes can move between them. All of these factors are taken into consideration when planning the nature of a preserve before the land is set aside.

In addition to the physical specifications of a preserve, there are a variety of regulations related to the use of a preserve. These can include anything from timber extraction, mineral extraction, regulated hunting, human habitation, and nondestructive human recreation. Many of the decisions to include these other uses are made based on political pressures rather than conservation considerations. On the other hand, in some cases, wildlife protection policies have been so strict that subsistence-living indigenous populations have been forced from ancestral lands that fell within a preserve. In other cases, even if a preserve is designed to protect wildlife, if the protections are not or cannot be enforced, the preserve status will have little meaning in the face of illegal poaching and timber extraction. This is a widespread problem with preserves in the tropics.

Climate change will create inevitable problems with the location of preserves as the species within them migrate to higher latitudes as the habitat of the preserve becomes less favorable. Planning for the effects of global warming on future preserves, or adding new preserves to accommodate the changes expected from global warming is in progress, but will only be as effective as the accuracy of the predictions of the effects of global warming on future habitats.

Finally, an argument can be made that conservation preserves reinforce the cultural perception that humans are separate from nature, can exist outside of it, and can only operate in ways that do damage to biodiversity. Creating preserves reduces the pressure on human activities outside the preserves to be sustainable and non-damaging to biodiversity. Ultimately, the political, economic, and human demographic pressures will degrade and reduce the size of conservation preserves if the activities outside them are not altered to be less damaging to biodiversity.

Process of Habitat Restoration

Habitat restoration is the process of bringing an area back to its natural state, before it was impacted through destructive human activities. It holds considerable promise as a mechanism for maintaining or restoring biodiversity. Reintroducing wolves, a top predator, to Yellowstone National Park in 1995 led to dramatic changes in the ecosystem that increased biodiversity. The wolves (Figure 6) function to suppress elk and coyote populations and provide more abundant resources to the detritivores. Reducing elk populations has allowed revegetation of riparian (the areas along the banks of a stream or river) areas, which has increased the diversity of species in that habitat. Reduction of coyote populations by wolves has increased the prey species previously suppressed by coyotes. In this habitat, the wolf is a keystone species, meaning a species that is instrumental in maintaining diversity within an ecosystem. Removing a keystone species from an ecological community causes a collapse in diversity. The results from the Yellowstone experiment suggest that restoring a keystone species effectively can have the effect of restoring biodiversity in the community. Ecologists have argued for the identification of keystone species where possible and for focusing protection efforts on these species. It makes sense to return the keystone species to the ecosystems where they have been removed.

Other large-scale restoration experiments underway involve dam removal. In the United States, since the mid-1980s, many aging dams are being considered for removal rather than replacement because of shifting beliefs about the ecological value of free-flowing rivers. The measured benefits of dam removal include restoration of naturally fluctuating water levels (often the purpose of dams is to reduce variation in river flows), which leads to increased fish diversity and improved water quality.

Figure 24. This photograph shows the Gibbon wolf pack in Yellowstone National Park, March 1, 2007. Wolves have been identified as a keystone species. (credit: Doug Smith, NPS)

In the Pacific Northwest of the United States, dam removal projects are expected to increase populations of salmon, which is considered a keystone species because it transports nutrients to inland ecosystems during its annual spawning migrations. In other regions, such as the Atlantic coast, dam removal has allowed the return of other spawning anadromous fish species (species that are born in fresh water, live most of their lives in salt water, and return to fresh water to spawn). Some of the largest dam removal projects have yet to occur or have happened too recently for the consequences to be measured, such as Elwha Dam on the Olympic Peninsula of Washington State. The large-scale ecological experiments that these removal projects constitute will provide valuable data for other dam projects slated either for removal or construction.

The Role of Zoos and Captive Breeding

Zoos have sought to play a role in conservation efforts both through captive breeding programs and education (Figure 25). The transformation of the missions of zoos from collection and exhibition facilities to organizations that are dedicated to conservation is ongoing. In general, it has been recognized that, except in some specific targeted cases, captive breeding programs for endangered species are inefficient and often prone to failure when the species are reintroduced to the wild. Zoo facilities are far too limited to contemplate captive breeding programs for the numbers of species that are now at risk. Education, on the other hand, is a potential positive impact of zoos on conservation efforts, particularly given the global trend to urbanization and the consequent reduction in contacts between people and wildlife. A number of studies have been performed to look at the effectiveness of zoos on people's attitudes and actions regarding conservation and at present, the results tend to be mixed.

Figure 25. Zoos and captive breeding programs help preserve many endangered species, such as this golden lion tamarin. (credit: Garrett Ziegler)

AT A GLANCE

Biodiversity exists at multiple levels of organization, and is measured in different ways depending on the goals of those taking the measurements. These include numbers of species, genetic diversity, chemical diversity, and ecosystem diversity. Humans use many compounds that were first discovered or derived from living organisms as medicines: secondary plant compounds, animal toxins, and antibiotics produced by bacteria and fungi. Ecosystems provide ecosystem services that support human agriculture: pollination, nutrient cycling, pest control, and soil development and maintenance. Loss of biodiversity threatens these ecosystem services and risks making food production more expensive or impossible. The core threats to biodiversity are human population growth and unsustainable resource use. Climate change is predicted to be a significant cause of extinction in the coming century. Exotic species have been the cause of a number of extinctions and are especially damaging to islands and lakes. International treaties such as CITES regulate the transportation of endangered species across international borders. In the United States, the Endangered Species Act protects listed species but is hampered by procedural difficulties and a focus on individual species. The Migratory Bird Act is an agreement between Canada and the United States to protect migratory birds. Presently, 11 percent of Earth's land surface is protected in some way. Habitat restoration has the potential to restore ecosystems to previous biodiversity levels before species become extinct. Examples of restoration include reintroduction of keystone species and removal of dams on rivers.

CHAPTER-6

EFFECTS OF ENVIRONMENTAL HAZARDS AND HUMAN HEALTH

6.1 THE IMPACTS OF ENVIRONMENTAL CONDITIONS ON PUBLIC HEALTH

Our industrialized society dumps huge amounts of pollutants and toxic wastes into the earth's biosphere without fully considering the consequences. Such actions seriously degrade the health of the earth's ecosystems, and this degradation ultimately affects the health and well-being of human populations.

For most of human history, **biological agents** were the most significant factor in health. These included pathogenic (disease causing) organisms such as bacteria, viruses, protozoa, and internal parasites. In modern times, cardiovascular diseases, cancer, and accidents are the leading killers in most parts of the world. However, infectious diseases still cause about 22 million deaths a year, mostly in undeveloped countries. These diseases include: tuberculosis, malaria, pneumonia, influenza, whooping cough, dysentery and Acquired Immune Deficiency Syndrome (AIDS). Most of those affected are children. Malnutrition, unclean water, poor sanitary conditions and lack of proper medical care all play roles in these deaths. Compounding the problems of infectious diseases are factors such as drug-resistant pathogens, insecticide resistant carriers, and overpopulation. Overuse of antibiotics have allowed pathogens to develop a resistance to drugs. For example, tuberculosis (TB) was nearly eliminated in most parts of the world, but drug-resistant strains have now reversed that trend. Another example is malaria. The insecticide DDT (a chemical called dichlorodiphenyltrichloroethane) was widely used to control malaria-carrying mosquito populations in tropical regions. However, after many years the mosquitoes developed a natural resistance to DDT and again spread the disease widely. Anti-malarial medicines were also over-prescribed, which allowed the malaria pathogen to become drug-resistant.

Chemical agents also have significant effects on human health. Toxic heavy metals, dioxins, pesticides, and endocrine disrupters are examples of these chemical agents. Heavy metals (e.g., mercury, lead, & cadmium) are typically produced as by-products of mining and manufacturing processes. All of them biomagnify (become more concentrated in species with increasing food chain level). For example, mercury from polluted water can accumulate in swordfish to levels toxic to humans. When toxic heavy metals get into the body, they accumulate in tissues and may eventually cause sickness or death. Studies show that people with above-average lead levels in their bones have an increased risk of developing attention deficit disorder and aggressive behavior. Lead can also damage brain cells and affect muscular coordination.

DDT PERSISTENT IN THE ENVIRONMENT

Figure 1. DDT was a commonly-used pesticide. "This work" by OpenStax is licensed under CC BY 4.0.

The pesticide DDT was widely used for decades. It was seen as an ideal pesticide because it is inexpensive and breaks down slowly in the environment. Unfortunately, the latter characteristic allows this chemical agent to biomagnify through the food chain. Populations of bird species at the top of the food chain, e.g., eagles and pelicans, are greatly affected by DDT in the environment. When these birds have sufficient levels of DDT, the shells of their eggs are so thin that they break, making reproduction impossible. After DDT was banned in the United States in 1972, affected bird populations made noticeable recoveries, including the iconic bald eagle.

6.2 ENVIRONMENTAL HEALTH

Environmental health is concerned with preventing disease, death and disability by reducing exposure to adverse environmental conditions and promoting behavioral change. It focuses on the direct an d indirect causes of diseases and injuries, and taps resources inside and outside the health care system to help improve health outcomes.

'Table 1. Typical Environmental Health Issues: Determinants and Health Consequences.

Underlying Determinants	Possible Adverse Health and Safety Consequences
Inadequate water (quantity and quality), sanitation (wastewater and excreta removal) and solid waste disposal, improper hygiene (hand washing)	Diarrheas and vector-related diseases, eg, malaria, schistosomiasis, dengue
Improper water resource management (urban and rural), including poor drainage	Vector-related diseases, eg, malaria, schistosomiasis
Crowded housing and poor ventilation of smoke	Acute and chronic respiratory diseases, including lung cancer (from coal and tobacco smoke inhalation)
Exposures to vehicular and industrial air pollution	Respiratory diseases, some cancers, and loss of IQ in children
Population movement and encroachment and construction, which affect feeding and breeding grounds of vectors, such as mosquitoes	Vector-related diseases, eg, malaria, schistosomiasis, and dengue fever, may also help spread other infectious diseases eg HIV/AIDS, Ebola fever
Exposure to naturally occurring toxic substances	Poisoning from, eg, arsenic, manganese, and fluorides
Natural resource degradation, eg, mudslides, poor drainage, erosion	Injury and death from mudslides and flooding
Climate change, partly from combustion of greenhouse gases in transportation, industry and poor energy conservation in housing, fuel, commerce, industry	Injury/death from: extreme heat/cold, storms, floods, fires. Indirect effects: spread of vector-borne diseases, aggravation of respiratory diseases, population dislocation, water pollution from sea level rise, etc.
Ozone depletion from industrial and commercial activity	Skin cancer, cataracts. Indirect effects: compromised food production, etc.

Poverty, Health and Environment

Environmental health risks can be grouped into two broad categories. Traditional hazards are related to poverty and the lack of development and mostly affect developing countries and poor people. Their impact exceeds that of modern health hazards by 10 times in Africa, 5 times in Asian countries (except for China), and 2.5 times in Latin America and Middle East (Figure 1). Water-related diseases caused by inadequate water supply and sanitation impose an especially large health burden in Africa, Asia, and the Pacific region. In India alone, over 700,000 children under 5 die annually from diarrhea. In Africa, malaria causes about 500,000 deaths annually. More than half of the world's households use unprocessed solid fuels, particularly biomass (crop residues, wood, and dung) for cooking and heating in inefficient stoves without proper ventilation, exposing people—mainly poor women and children—to high levels of indoor air pollution(IAP). IAP causes about 2 million deaths in each year.

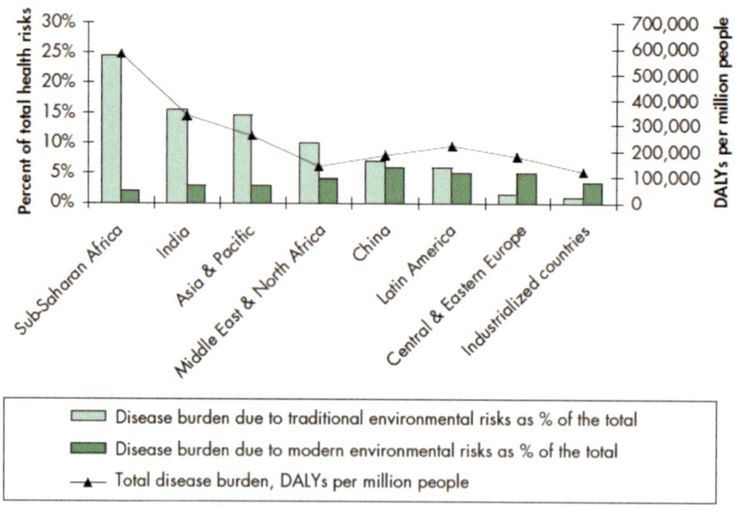

Figure 1. Traditional environmental health hazards prevail in developing countries but modern risks are also significant.

Modern hazards, caused by technological development, prevail in industrialized countries where exposure to traditional hazards is low. The contribution of modern environmental risks to the disease burden in most developing countries is similar to – and in quite a few countries, greater than – that in rich countries. Urban air pollution, for example, is highest in parts of China, India and some cities in Asia and Latin America. Poor people increasingly experience a "double burden" of traditional and modern environmental health risks. Their total burden of illness and death from all causes per million people is about twice that in rich countries, and the disease burden from environmental risks is 10 times greater.

Environmental Health and Child Survival

Figure 2. Malnourished children in Niger, during the 2005 famine.

Worldwide, the top killers of children under five are acute respiratory infections (from indoor air pollution); diarrheal diseases (mostly from poor water, sanitation, and hygiene); and infectious diseases such as malaria. Children are especially susceptible to environmental factors that put them at risk of developing illness early in life. Malnutrition (the condition that occurs when body does not get enough nutrients) is an important contributor to child mortality—malnutrition and environmental infections are inextricably linked. The World Health Organization (WHO) recently concluded that about 50% of the consequences of malnutrition are in fact caused by inadequate water and sanitation provision and poor hygienic practices.

Limited access to water and sanitation

With 1.1 billion people lacking access to safe drinking water and 2.6 billion without adequate sanitation, the magnitude of the water and sanitation problem remains significant. Each year contaminated water and poor sanitation contribute to 5.4 billion cases of diarrhea worldwide and 1.6 million deaths, mostly among children under the age of five. Intestinal worms, which thrive in poor sanitary conditions, infect close to 90 percent of children in the developing world and, depending on the severity of the infection may lead to malnutrition, anemia, or stunted growth. About 6 million people are blind from trachoma, a disease caused by the lack of clean water combined with poor hygiene practices.

Indoor Air Pollution

Indoor air pollution—a much less publicized source of poor health—is responsible for more than 1.6 million deaths per year and for 2.7% of global burden of disease. It is estimated that half of the world's population, mainly in developing countries, uses solid fuels (biomass and coal) for household cooking and space heating. Cooking and heating with such solid fuels on open fires or stoves without chimneys lead to indoor air pollution and subsequently, respiratory infections. Exposure to these health-damaging pollutants is particularly high among women and children in developing countries, who spend the most time inside the household. As many as half of the deaths attributable to indoor use of solid fuel are of children under the age of five.

Malaria

Approximately 40% of the world's people—mostly those living in the world's poorest countries—are at risk from malaria. Malaria is an infectious disease spread by mosquitoes but caused by a single-celled parasite called Plasmodium. Every year, more than 200 million people become infected with malaria and about 430,000 die, with most cases and deaths found in Sub-Saharan Africa. However, Asia, Latin America, the Middle East, and parts of Europe are also affected. Pregnant women are especially at high risk of malaria. Non-immune pregnant women risk both acute and severe clinical disease, resulting in fetal loss in up to 60% of such women and maternal deaths in more than 10%, including a 50% mortality rate for those with severe disease. Semi-immune pregnant women with malaria infection risk severe anemia and impaired fetal growth, even if they show no signs of acute clinical disease. An estimated 10,000 women and 200,000 infants die annually as a result of malaria infection during pregnancy.

Important Emerging Diseases

Emerging and re-emerging diseases have been defined as infectious diseases of humans whose occurrence during the past two decades has substantially increased or threatens to increase in the near future relative to populations affected, geographic distribution, or magnitude of impacts. Examples include Ebola virus, West Nile virus, Zika virus, sudden acute respiratory syndrome (SARS), H1N1 influenza; swine and avian influenza (swine, bird flu), HIV, and a variety of other viral, bacterial, and protozoal diseases.

A variety of environmental factors may contribute to re-emergence of a particular disease, including temperature, moisture, human food or animal feed sources, etc. Disease re-emergence may be caused by the coincidence of several of these environmental and/or social factors to allow optimal conditions for transmission of the disease.

Ebola, previously known as Ebola hemorrhagic fever, is a rare and deadly disease caused by infection with one of the Ebola virus strains. Ebola can cause disease in humans and nonhuman primates. The 2014 Ebola epidemic is the largest in history (with over 28,000 cases and 11,302 deaths), affecting multiple countries in West Africa. There were a small number of cases reported in Nigeria and Mali and a single case reported in Senegal; however, these cases were contained, with no further spread in these countries.

The HIV/AIDS epidemic has spread with ferocious speed. Virtually unknown 20 years ago, HIV has infected more than 60 million people worldwide. Each day, approximately 14,000 new infections occur, more than half of them among young people below age 25. Over 95 percent of PLWHA (People Living With HIV/AIDS) are in low- and middle- income countries. More than 20 million have died from AIDS, over 3 million in 2002 alone. AIDS is now the leading cause of death in Sub-Saharan Africa and the fourth-biggest killer globally. The epidemic has cut life expectancy by more than 10 years in several nations.

It seems likely that a wide variety of infectious diseases have affected human populations for thousands of years emerging when the environmental, host, and agent conditions were favorable. Expanding human populations have increased the potential for transmission of infectious disease as a result of close human proximity and increased likelihood for humans to be in "the wrong place at the right time" for disease to occur (eg, natural disasters or political conflicts). Global travel increases the potential for a carrier of disease to transmit infection thousands of miles away in just a few hours, as evidenced by WHO precautions concerning international travel and health.

Antibiotic Resistance

Antibiotics and similar drugs, together called antimicrobial agents, have been used for the last 70 years to treat patients who have infectious diseases. Since the 1940s, these drugs have greatly reduced illness and death from infectious diseases. However, these drugs have been used so widely and for so long that the infectious organisms the antibiotics are designed to kill have adapted to them, making the drugs less effective. Antibiotic resistance occurs when bacteria change in a way that reduces the effectiveness of drugs, chemicals, or other agents designed to cure or prevent infections. This is caused by the process of evolution through natural selection (Figure 3). The antibiotic-resistant bacteria survive and continue to multiply, causing more harm.

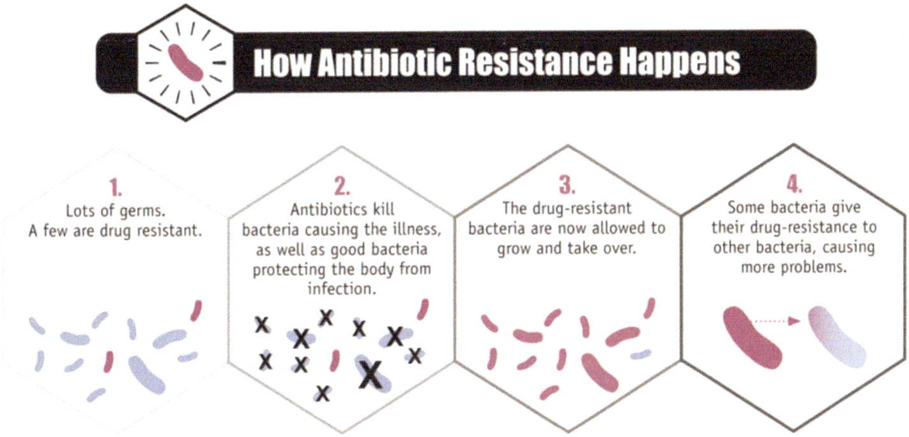

Figure 3. Antibiotic resistance is of increasing concern to medical professionals.

New forms of antibiotic resistance can cross international boundaries and spread between continents with ease. Many forms of resistance spread with remarkable speed. Each year in the United States, at least 2 million people acquire serious infections with bacteria that are resistant to one or more of the antibiotics designed to treat those infections. At least 23,000 people die each year in the US as a direct result of these antibiotic-resistant infections. Many more die from other conditions that were complicated by an antibiotic-resistant infection. The use of antibiotics is the single most important factor leading to antibiotic resistance around the world.

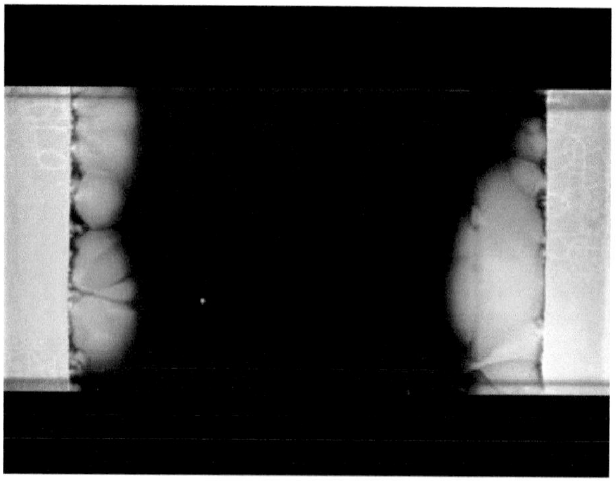

Figure 4. Harvard Medical school, where they show bacteria adapting very quickly to apparently deadly conditions.

Antibiotics are among the most commonly prescribed drugs used in human medicine, but up to 50% of all the antibiotics prescribed for people are not needed or are not optimally effective as prescribed.

During recent years, there has been growing concern over methicillin-resistant *Staphylococcus aureus* **(MRSA),** a bacterium that is resistant to many antibiotics. In the community, most MRSA infections are skin infections. In medical facilities, MRSA causes life-threatening bloodstream infections, pneumonia and surgical site infections.

6.3 ENVIRONMENTAL TOXICOLOGY: HUMANS AND ENVIRONMENT

Environmental toxicology is the scientific study of the health effects associated with exposure to toxic chemicals (Table 1) occurring in the natural, work, and living environments. The term also describes the management of environmental toxins and toxicity, and the development of protections for humans and the environment.

The table below lists top 20 substances, in order of priority, which are determined to pose the most significant potential threat to human health. This priority list is not a list of "most toxic" substances but rather a prioritization of substances based on a combination of their frequency, toxicity, and potential for human exposure at various sites.

Table 1. The ATSDR 2013 Substance Priority List.

2013 RANK	NAME
1	ARSENIC
2	LEAD
3	MERCURY
4	VINYL CHLORIDE
5	POLYCHLORINATED BIPHENYLS
6	BENZENE
7	CADMIUM
8	BENZO(A)PYRENE
9	POLYCYCLIC AROMATIC HYDROCARBONS
10	BENZO(B)FLUORANTHENE
11	CHLOROFORM
12	AROCLOR 1260
13	DDT, P,P'-
14	AROCLOR 1254
15	DIBENZO(A,H)ANTHRACENE
16	TRICHLOROETHYLENE
17	CHROMIUM, HEXAVALENT
18	DIELDRIN
19	PHOSPHORUS, WHITE
20	HEXACHLOROBUTADIENE

Routes of Exposure to Chemicals

In order to cause health problems, chemicals must enter your body. There are three main "routes of exposure," or ways a chemical can get into your body.

Breathing (inhalation): Breathing in chemical gases, mists, or dusts that are in the air.

Skin or eye contact: Getting chemicals on the skin, or in the eyes. They can damage the skin, or be absorbed through the skin into the bloodstream.

Swallowing (ingestion): This can happen when chemicals have spilled or settled onto food, beverages, cigarettes, beards, or hands.

Once chemicals have entered your body, some can move into your bloodstream and reach internal "target" organs, such as the lungs, liver, kidneys, or nervous system.

What Forms do Chemicals Take?

Chemical substances can take a variety of forms. They can be in the form of solids, liquids, dusts, vapors, gases, fibers, mists and fumes. The form a substance is in has a lot to do with how it gets into your body and what harm it can cause. A chemical can also change forms. For example, liquid solvents can evaporate and give off vapors that you can inhale. Sometimes chemicals are in a form that can't be seen or smelled, so they can't be easily detected.

What Health Effects Can Chemicals Cause?

An **acute effect** of a contaminant (The term "**contaminant**" means hazardous substances, pollutants, pollution, and chemicals) is one that occurs rapidly after exposure to a large amount of that substance. A chronic effect of a contaminant results from exposure to small amounts of a substance over a long period of time. In such a case, the effect may not be immediately obvious. **Chronic effect** are difficult to measure, as the effects may not be seen for years. Long-term exposure to cigarette smoking, low level radiation exposure, and moderate alcohol use are all thought to produce chronic effects. For centuries, scientists have known that just about any substance is toxic in sufficient quantities. For example, small amounts of selenium are required by living organisms for proper functioning, but large amounts may cause cancer. The effect of a certain chemical on an individual depends on the dose (amount) of the chemical. This relationship is often illustrated by a dose-response curve which shows the relationship between dose and the response of the individual. Lethal doses in humans have been determined for many substances from information gathered from records of homicides, accidental poisonings, and testing on animals.

A dose that is lethal to 50% of a population of test animals is called the lethal dose-50% or LD-50. Determination of the LD-50 is required for new synthetic chemicals in order to give a measure of their toxicity. A dose that causes 50% of a population to exhibit any significant response (e.g., hair loss, stunted development) is referred to as the effective dose-50% or ED-50. Some toxins have a threshold amount below which there is no apparent effect on the exposed population.

Environmental Contaminants: An Overview

The contamination of the air, water, or soil with potentially harmful substances can affect any person or community. Contaminants (Table 2) are often chemicals found in the environment in amounts higher than what would be there naturally. We can be exposed to these contaminants from a variety of residential, commercial, and industrial sources. Sometimes harmful environmental contaminants occur biologically, such as mold or a toxic algae bloom.

Table 2. Classification of Environmental Contaminants

Contaminant	Definition
Carcinogen	An agent which may produce cancer (uncontrolled cell growth), either by itself or in conjunction with another substance. Examples include formaldehyde, asbestos, radon, vinyl chloride, and tobacco.
Teratogen	A substance which can cause physical defects in a developing embryo. Examples include alcohol and cigarette smoke.
Mutagen	A material that induces genetic changes (mutations) in the DNA. Examples include radioactive substances, x-rays and ultraviolet radiation.
Neurotoxicant	A substance that can cause an adverse effect on the chemistry, structure or function of the nervous system. Examples include lead and mercury.
Endocrine disruptor	A chemical that may interfere with the body's endocrine (hormonal) system and produce adverse developmental, reproductive, neurological, and immune effects in both humans and wildlife. A wide range of substances, both natural and man-made, are thought to cause endocrine disruption, including pharmaceuticals, dioxin and dioxin-like compounds, arsenic, polychlorinated biphenyls (PCBs), DDT and other pesticides, and plasticizers such as bisphenol A (BPA).

Some Common Contaminants

Arsenic is a naturally occurring element that is normally present throughout our environment in water, soil, dust, air, and food. Levels of arsenic can regionally vary due to farming and industrial activity as well as natural geological processes. The arsenic from farming and smelting tends to bind strongly to soil and is expected to remain near the surface of the land for hundreds of years as a long-term source of exposure. Wood that has been treated with chromated copper arsenate (CCA) is commonly found in decks and railing in existing homes and outdoor structures such as playground equipment. Some underground aquifers are located in rock or soil that has naturally high arsenic content.

Most arsenic gets into the body through ingestion of food or water. Arsenic in drinking water is a problem in many countries around the world, including Bangladesh, Chile, China, Vietnam, Taiwan, India, and the United States. Arsenic may also be found in foods, including rice and some fish, where it is present due to uptake from soil and water. It can also enter the body by breathing dust containing arsenic. Researchers are finding that arsenic, even at low levels, can interfere with the body's endocrine system. Arsenic is also a known human carcinogen associated with skin, lung, bladder, kidney, and liver cancer.

Mercury is a naturally occurring metal, a useful chemical in some products, and a potential health risk. Mercury exists in several forms; the types people are usually exposed to are methylmercury and elemental mercury. Elemental mercury at room temperature is a shiny, silver-white liquid which can produce a harmful odorless vapor. Methylmercury, an organic compound, can build up in the bodies of long-living, predatory fish. To keep mercury out of the fish we eat and the air we breathe, it's important to take mercury-containing products to a hazardous waste facility for disposal. Common products sold today that contain small amounts of mercury include fluorescent lights and button-cell batteries.

Although fish and shellfish have many nutritional benefits, consuming large quantities of fish increases a person's exposure to mercury. Pregnant women who eat fish high in mercury on a regular basis run the risk of permanently damaging their developing fetuses. Children born to these mothers may exhibit motor difficulties, sensory problems and cognitive deficits. Figure 1 identifies the typical (average) amounts of mercury in commonly consumed commercial and sport-caught fish.

Bisphenol A (BPA) is a chemical synthesized in large quantities for use primarily in the production of polycarbonate plastics and epoxy resins. Polycarbonate plastics have many applications including use in some food and drink packaging, e.g., water and infant bottles, compact discs, impact-resistant safety equipment, and medical devices. Epoxy resins are used as lacquers to coat metal products such as food cans, bottle tops, and water supply pipes. Some dental sealants and composites may also contribute to BPA exposure. The primary source of exposure to BPA for most people is through the diet. Bisphenol A can leach into food from the protective internal epoxy resin coatings of canned foods and from consumer products such as polycarbonate tableware, food storage containers, water bottles, and baby bottles. The degree to which BPA leaches from polycarbonate bottles into liquid may depend more on the temperature of the liquid or bottle, than the age of the container. BPA can also be found in breast milk.

Figure 5. Mercury concentrations in fish can reach potentially dangerous levels (published by the Maine Center for Disease Control & Prevention).

What can I do to prevent exposure to BPA?

Some animal studies suggest that infants and children may be the most vulnerable to the effects of BPA. Parents and caregivers, can make the personal choice to reduce exposures of their infants and children to BPA:

- Don't microwave polycarbonate plastic food containers. Polycarbonate is strong and durable, but over time it may break down from over use at high temperatures.
- Plastic containers have recycle codes on the bottom. Some, but not all, plastics that are marked with recycle codes 3 or 7 may be made with BPA.
- Reduce your use of canned foods.
- When possible, opt for glass, porcelain or stainless steel containers, particularly for hot food or liquids.
- Use baby bottles that are BPA free.

Phthalates are a group of synthetic chemicals used to soften and increase the flexibility of plastic and vinyl. Polyvinyl chloride is made softer and more flexible by the addition of phthalates. Phthalates are used in hundreds of consumer products. Phthalates are used in cosmetics and personal care products, including perfume, hair spray, soap, shampoo, nail polish, and skin moisturizers. They are used in consumer products such as flexible plastic and vinyl toys, shower curtains, wallpaper, vinyl miniblinds, food packaging, and plastic wrap. Exposure to low levels of phthalates may come from eating food packaged in plastic that contains phthalates or breathing dust in rooms with vinyl miniblinds, wallpaper, or recently installed flooring that contain phthalates. We can be exposed to phthalates by drinking water that contains phthalates. Phthalates are suspected to be endocrine disruptors.

Lead is a metal that occurs naturally in the rocks and soil of the earth's crust. It is also produced from burning fossil fuels such as coal, oil, gasoline, and natural gas; mining; and manufacturing. Lead has no distinctive taste or smell. The chemical symbol for elemental lead is Pb. Lead is used to produce batteries, pipes, roofing, scientific electronic equipment, military tracking systems, medical devices, and products to shield X-rays and nuclear radiation. It is used in ceramic glazes and crystal glassware. Because of health concerns, lead and lead compounds were banned from house paint in 1978; from solder used on water pipes in 1986; from gasoline in 1995; from solder used on food cans in 1996; and from tin-coated foil on wine bottles in 1996. The U.S. Food and Drug Administration has set a limit on the amount of lead that can be used in ceramics.

Lead and lead compounds are listed as "reasonably anticipated to be a human carcinogen". It can affect almost every organ and system in your body. It can be equally harmful if breathed or swallowed. The part of the body most sensitive to lead exposure is the central nervous system, especially in children, who are more vulnerable to lead poisoning than adults. A child who swallows large amounts of lead can develop brain damage that can cause convulsions and death; the child can also develop blood anemia, kidney damage, colic, and muscle weakness. Repeated low levels of exposure to lead can alter a child's normal mental and physical growth and result in learning or behavioral problems. Exposure to high levels of lead for pregnant women can cause miscarriage, premature births, and smaller babies. Repeated or chronic exposure can cause lead to accumulate in your body, leading to lead poisoning.

Figure 6. Nail products are known to contain toxic chemicals, such as dibutyl phthalate (DBP), toluene, and formaldehyde.

Formaldehyde is a colorless, flammable gas or liquid that has a pungent, suffocating odor. It is a volatile organic compound, which is an organic compound that easily becomes a vapor or gas. It is also naturally produced in small, harmless amounts in the human body. The primary way we can be exposed to formaldehyde is by breathing air containing it. Releases of formaldehyde into the air occur from industries using or manufacturing formaldehyde, wood products (such as particle-board, plywood, and furniture), automobile exhaust, cigarette smoke, paints and varnishes, and carpets and permanent press fabrics. Nail polish, and commercially applied floor finish emit formaldehyde.

In general, indoor environments consistently have higher concentrations than outdoor environments, because many building materials, consumer products, and fabrics emit formaldehyde. Levels of formaldehyde measured in indoor air range from 0.02–4 parts per million (ppm). Formaldehyde levels in outdoor air range from 0.001 to 0.02 ppm in urban areas.

Radiation

Radiation is energy given off by atoms and is all around us. We are exposed to radiation every day from natural sources like soil, rocks, and the sun. We are also exposed to radiation from man-made sources like medical X-rays and smoke detectors. We're even exposed to low levels of radiation on cross-country flights, from watching television, and even from some construction materials. You cannot see, smell or taste radiation. Some types of radioactive materials are more dangerous than others. So it's important to carefully manage radiation and radioactive substances to protect health and the environment.

Radon is a radioactive gas that is naturally-occurring, colorless, and odorless. It comes from the natural decay of uranium or thorium found in nearly all soils. It typically moves up through the ground and into the home through cracks in floors, walls and foundations. It can also be released from building materials or from well water. Radon breaks down quickly, giving off radioactive particles. Long-term exposure to these particles can lead to lung cancer. Radon is the leading cause of lung cancer among nonsmokers, according to the U.S. Environmental Protection Agency, and the second leading cause behind smoking.

6.4 BIOREMEDIATION PROCESS

Bioremediation is a waste management technique that involves the use of organisms such as plants, bacteria, and fungi to remove or neutralize pollutants from a contaminated site. According to the United States EPA, bioremediation is a "treatment that uses naturally occurring organisms to break down hazardous substances into less toxic or non toxic substances".

Bioremediation is widely used to treat human sewage and has also been used to remove agricultural chemicals (pesticides and fertilizers) that leach from soil into groundwater. Certain toxic metals, such as selenium and arsenic compounds, can also be removed from water by bioremediation. Mercury is an example of a toxic metal that can be removed from an environment by bioremediation. Mercury is an active ingredient of some pesticides and is also a byproduct of certain industries, such as battery production. Mercury is usually present in very low concentrations in natural environments but it is highly toxic because it accumulates in living tissues. Several species of bacteria can carry out the biotransformation of toxic mercury into nontoxic forms. These bacteria, such as *Pseudomonas aeruginosa*, can convert Hg^{2+} to Hg, which is less toxic to humans.

Probably one of the most useful and interesting examples of the use of prokaryotes for bioremediation purposes is the cleanup of oil spills. The importance of prokaryotes to petroleum bioremediation has been demonstrated in several oil spills in recent years, such as the Exxon Valdez spill in Alaska (1989) (Figure 1), the Prestige oil spill in Spain (2002), the spill into the Mediterranean from a Lebanon power plant (2006,) and more recently, the BP oil spill in the Gulf of Mexico (2010). To clean up these spills, bioremediation is promoted by adding inorganic nutrients that help bacteria already present in the environment to grow. Hydrocarbon-degrading bacteria feed on the hydrocarbons in the oil droplet, breaking them into inorganic compounds. Some species, such as *Alcanivorax borkumensis*, produce surfactants that solubilize the oil, while other bacteria degrade the oil into carbon dioxide. In the case of oil spills in the ocean, ongoing, natural bioremediation tends to occur, inasmuch as there are oil-consuming bacteria in the ocean prior to the spill. Under ideal conditions, it has been reported that up to 80 percent of the nonvolatile components in oil can be degraded within 1 year of the spill. Researchers have genetically engineered other bacteria to consume petroleum products; indeed, the first patent application for a bioremediation application in the U.S. was for a genetically modified oil-eating bacterium.

Figure 7. Figure 1. (a) Cleaning up oil after the Valdez spill in Alaska, the workers hosed oil from beaches and then used a floating boom to corral the oil, which was finally skimmed from the water surface. Some species of bacteria are able to solubilize and degrade the oil. (b) One of the most catastrophic consequences of oil spills is the damage to fauna. (credit a: modification of work by NOAA; credit b: modification of work by GOLUBENKOV, NGO: Saving Taman)

There are a number of cost/efficiency advantages to bioremediation, which can be employed in areas that are inaccessible without excavation. For example, hydrocarbon spills (specifically, oil spills) or certain chlorinated solvents may contaminate groundwater, which can be easier to treat using bioremediation than more conventional approaches. This is typically much less expensive than excavation followed by disposal elsewhere, incineration, or other off-site treatment strategies. It also reduces or eliminates the need for "pump and treat", a practice common at sites where hydrocarbons have contaminated clean groundwater. Using prokaryotes for bioremediation of hydrocarbons also has the advantage of breaking down contaminants at the molecular level, as opposed to simply chemically dispersing the contaminant.

Case Study: The Love Canal Disaster

One of the most famous and important examples of groundwater pollution in the U.S. is the **Love Canal tragedy** in Niagara Falls, New York. It is important because the pollution disaster at Love Canal, along with similar pollution calamities at that time (Times Beach, Missouri and Valley of Drums, Kentucky), helped to create **Superfund**, a federal program instituted in 1980 and designed to identify and clean up the worst of the hazardous chemical waste sites in the U.S.

Figure 8. Love Canal. Source: US Environmental Protection Agency

Love Canal is a neighborhood in Niagara Falls named after a large ditch (approximately 15 m wide, 3–12 m deep, and 1600 m long) that was dug in the 1890s for hydroelectric power. The ditch was abandoned before it actually generated any power and went mostly unused for decades, except for swimming by local residents. In the 1920s Niagara Falls began dumping urban waste into Love Canal, and in the 1940s the U.S. Army dumped waste from World War II there, including waste from the frantic effort to build a nuclear bomb. Hooker Chemical purchased the land in 1942 and lined it with clay. Then, the company put into Love Canal an estimated 21,000 tons of hazardous chemical waste, including the carcinogens benzene, dioxin, and PCBs in large metal barrels and covered them with more clay. In 1953, Hooker sold the land to the Niagara Falls school board for $1, and included a clause in the sales contract that both described the land use (filled with chemical waste) and absolved them from any future damage claims from the buried waste. The school board promptly built a public school on the site and sold the surrounding land for a housing project that built 200 or so homes along the canal banks and another 1,000 in the neighborhood (Figure 1). During construction, the canal's clay cap and walls were breached, damaging some of the metal barrels.

Eventually, the chemical waste seeped into people's basements, and the metal barrels worked their way to the surface. Trees and gardens began to die; bicycle tires and the rubber soles of children's shoes disintegrated in noxious puddles. From the 1950s to the late 1970s, residents repeatedly complained of strange odors and substances that surfaced in their yards. City officials investigated the area, but did not act to solve the problem. Local residents allegedly experienced major health problems including high rates of miscarriages, birth defects, and chromosome damage, but studies by the New York State Health Department disputed that. Finally, in 1978 President Carter declared a state of emergency at Love Canal, making it the first human-caused environmental problem to be designated that way. The Love Canal incident became a symbol of improperly stored chemical waste. Clean up of Love Canal, which was funded by Superfund and completely finished in 2004, involved removing contaminated soil, installing drainage pipes to capture contaminated groundwater for treatment, and covering it with clay and plastic. In 1995, Occidental Chemical (the modern name for Hooker Chemical) paid $102 million to Superfund for cleanup and $27 million to Federal Emergency Management Association for the relocation of more than 1,000 families. New York State paid $98 million to EPA and the US government paid $8 million for pollution by the Army. The total clean up cost was estimated to be $275 million.

The Love Canal tragedy helped to create Superfund, which has analyzed tens of thousands of hazardous waste sites in the U.S. and cleaned up hundreds of the worst ones. Nevertheless, over 1,000 major hazardous waste sites with a significant risk to human health or the environment are still in the process of being cleaned.

AT A GLANCE

Environmental health is concerned with preventing disease, death and disability by reducing exposure to adverse environmental conditions and promoting behavioral change. It focuses on the direct and indirect causes of diseases and injuries, and taps resources inside and outside the health care system to help improve health outcomes. Environmental health risks can be grouped into two broad categories. Traditional hazards related to poverty and lack of development affect developing countries and poor people most. Modern hazards, caused by development that lacks environmental safeguards, such as urban (outdoor) air pollution and exposure to agro-industrial chemicals and waste, prevail in industrialized countries, where exposure to traditional hazards is low. Each year contaminated water and poor sanitation contribute to 5.4 billion cases of diarrhea worldwide and 1.6 million deaths, mostly among children under the age of five. Indoor air pollution—a much less publicized source of poor health—is responsible for more than 1.6 million deaths per year and for 2.7 percent of global burden of disease.

Emerging and reemerging diseases have been defined as infectious diseases of humans whose occurrence during the past two decades has substantially increased or threatens to increase in the near future relative to populations affected, geographic distribution, or magnitude of impacts. Antibiotic resistance is a global problem. New forms of antibiotic resistance can cross international boundaries and spread between continents. Environmental toxicology is the scientific study of the health effects associated with exposure to toxic chemicals and systems occurring in the natural, work, and living environments; the management of environmental toxins and toxicity; and the development of protections for humans, animals, and plants. Environmental contaminants are chemicals found in the environment in amounts higher than what would be there naturally. We can be exposed to these contaminants from a variety of residential, commercial, and industrial sources.

CHAPTER-7

WATER SUPPLY, AVAILABILITY AND USE

7.1 WATER CYCLE AND SUPPLY OF FRESH WATER

Water, air, and food are the most important natural resources to people. Humans can live only a few minutes without oxygen, less than a week without water, and about a month without food. Water also is essential for our oxygen and food supply. Plants breakdown water and use it to create oxygen during the process of photosynthesis.

Water is the most essential compound for all living things. Human babies are approximately 75% water and adults are 60% water. Our brain is about 85% water, blood and kidneys are 83% water, muscles are 76% water, and even bones are 22% water. We constantly lose water by perspiration; in temperate climates we should drink about 2 quarts of water per day and people in hot desert climates should drink up to 10 quarts of water per day. Loss of 15% of body-water usually causes death.

Earth is truly the Water Planet. The abundance of liquid water on Earth's surface distinguishes us from other bodies in the solar system. About 70% of Earth's surface is covered by oceans and approximately half of Earth's surface is obscured by clouds (also made of water) at any time. There is a very large volume of water on our planet, about 1.4 billion cubic kilometers (km3) (330 million cubic miles) or about 53 billion gallons per person on Earth. All of Earth's water could cover the United States to a depth of 145 km (90 mi). From a human perspective, the problem is that over 97% of it is seawater, which is too salty to drink or use for irrigation. The most commonly used water sources are rivers and lakes, which contain less than 0.01% of the world's water!

One of the most important environmental goals is to provide clean water to all people. Fortunately, water is a renewable resource and is difficult to destroy. Evaporation and precipitation combine to replenish our fresh water supply constantly; however, water availability is complicated by its uneven distribution over the Earth. Arid climate and densely populated areas have combined in many parts of the world to create water shortages, which are projected to worsen in the coming years due to population growth and climate change. Human activities such as water overuse and water pollution have compounded significantly the water crisis that exists today. Hundreds of millions of people lack access to safe drinking water, and billions of people lack access to improved sanitation as simple as a pit latrine. As a result, nearly two million people die every year from diarrheal diseases and 90% of those deaths occur among children under the age of 5. Most of these are easily prevented deaths.

Water Reservoirs and Water Cycle

Water is the only common substance that occurs naturally on earth in three forms: solid, liquid and gas. It is distributed in various locations, called water reservoirs. The oceans are by far the largest of the reservoirs with about 97% of all water but that water is too saline for most human uses (Figure 1). Ice caps and glaciers are the largest reservoirs of fresh water but this water is inconveniently located, mostly in Antarctica and Greenland. Shallow groundwater is the largest reservoir of usable fresh water. Although rivers and lakes are the most heavily used water resources, they represent only a tiny amount of the world's water. If all of world's water was shrunk to the size of 1 gallon, then the total amount of fresh water would be about 1/3 cup, and the amount of readily usable fresh water would be 2 tablespoons.

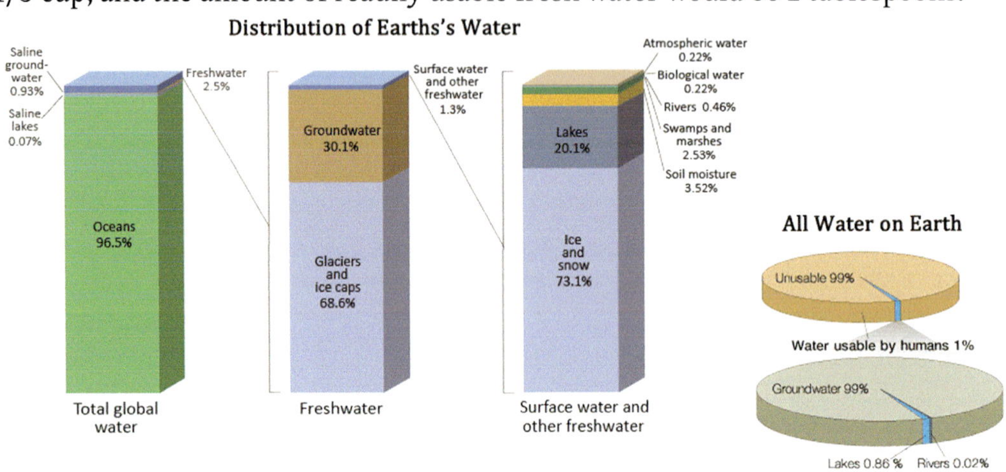

Figure 1. Earth's Water Reservoirs. Bar chart Distribution of Earth's water including total global water, fresh water, and surface water and other fresh water and Pie chart Water usable by humans and sources of usable water. Source: United States Geographical Survey Igor Skiklomanov's chapter "World fresh water resources" in Peter H. Gleick (editor), 1993, Water in Crisis: A Guide to the World's Fresh Water Resources

The water (or hydrologic) cycle (that was covered in Chapter 3.2) shows the movement of water through different reservoirs, which include oceans, atmosphere, glaciers, groundwater, lakes, rivers, and biosphere. Solar energy and gravity drive the motion of water in the water cycle. Simply put, the water cycle involves water moving from oceans, rivers, and lakes to the atmosphere by evaporation, forming clouds. From clouds, it falls as precipitation (rain and snow) on both water and land. The water on land can either return to the ocean by surface runoff, rivers, glaciers, and subsurface groundwater flow, or return to the atmosphere by evaporation or transpiration (loss of water by plants to the atmosphere).

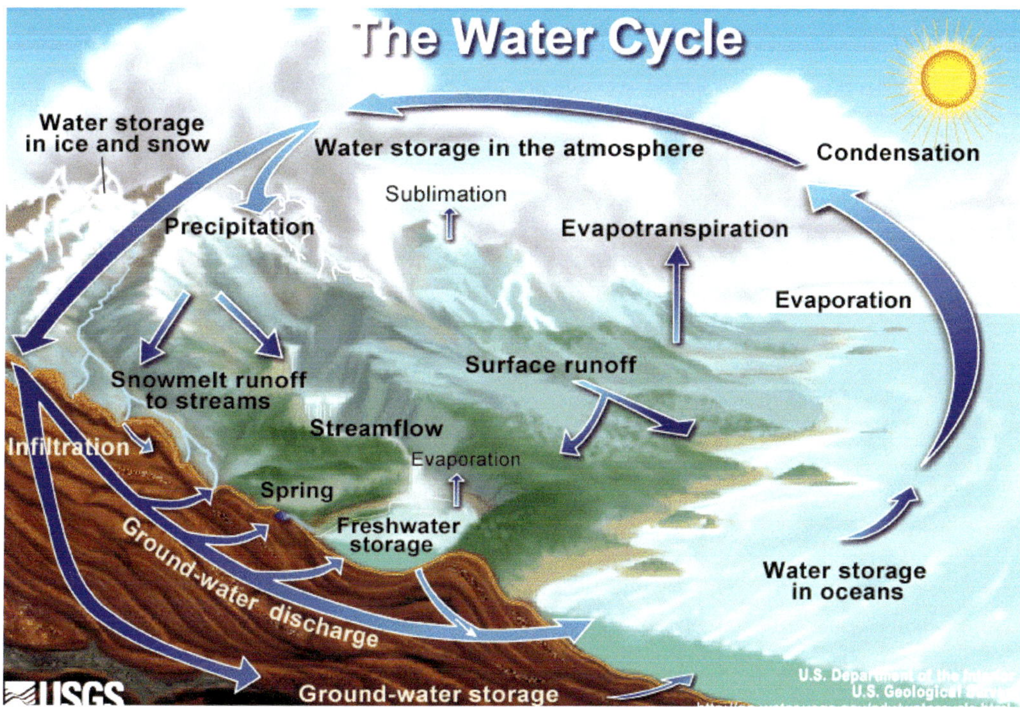

Figure 2. The Water Cycle. Arrows depict movement of water to different reservoirs located above, at, and below Earth's surface. Source: United States Geological Survey

An important part of the water cycle is how water varies in salinity, which is the abundance of dissolved ions in water. The saltwater in the oceans is highly saline, with about 35,000 mg of dissolved ions per liter of seawater. **Evaporation** (where water changes from liquid to gas at ambient temperatures) is a distillation process that produces nearly pure water with almost no dissolved ions. As water vaporizes, it leaves the dissolved ions in the original liquid phase. Eventually, **condensation** (where water changes from gas to liquid) forms clouds and sometimes precipitation (rain and snow). After rainwater falls onto land, it dissolves minerals in rock and soil, which increases its salinity. Most lakes, rivers, and near-surface groundwater have a relatively low salinity and are called freshwater. The next several sections discuss important parts of the water cycle relative to fresh water resources.

Precipitation: Primary Fresh Water Resources

Precipitation levels are unevenly distributed around the globe, affecting fresh water availability (Figure 3). More precipitation falls near the equator, whereas less precipitation tends to fall near 30 degrees north and south latitude, where the world's largest deserts are located. These rainfall and climate patterns are related to global wind circulation cells. The intense sunlight at the equator heats air, causing it to rise and cool, which decreases the ability of the air mass to hold water vapor and results in frequent rainstorms. Around 30 degrees north and south latitude, descending air conditions produce warmer air, which increases its ability to hold water vapor and results in dry conditions. Both the dry air conditions and the warm temperatures of these latitude belts favor evaporation. Global precipitation and climate patterns are also affected by the size of continents, major ocean currents, and mountains.

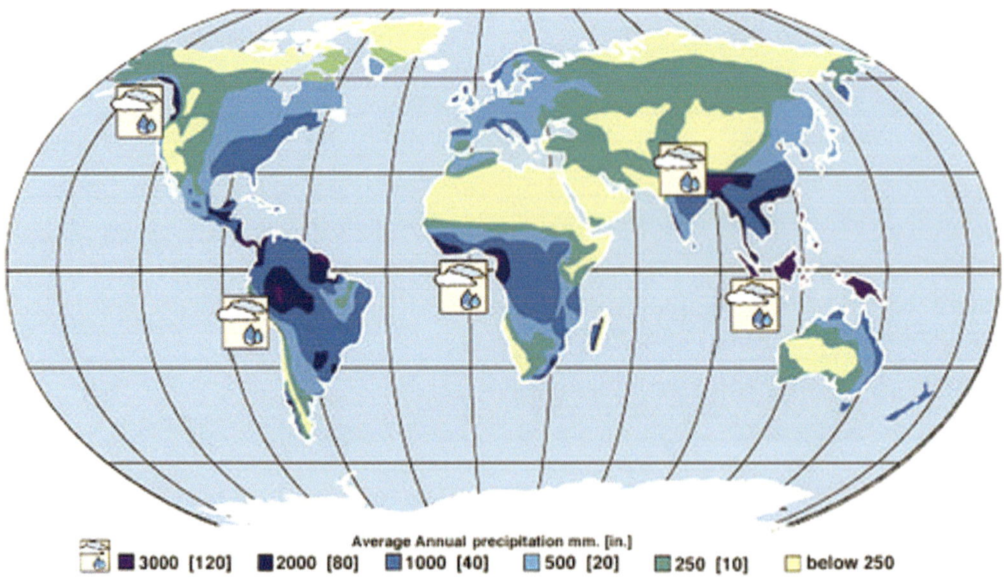

Figure 3. World Rainfall Map. The false-color map above shows the amount of rain that falls around the world. Areas of high rainfall include Central and South America, western Africa, and Southeast Asia. Since these areas receive so much rainfall, they are where most of the world's rainforests grow. Areas with very little rainfall usually turn into deserts. The desert areas include North Africa, the Middle East, western North America, and Central Asia. Source: United States Geological Survey Earth Forum, Houston Museum Natural Science

Rivers, Lakes and Glaciers: Surface Water Resources

Figure 4. Surface Runoff Surface runoff, part of overland flow in the water cycle Source: James M. Pease at Wikimedia Commons

Flowing water from rain and melted snow on land enters river channels by surface runoff (Figure 4) and groundwater seepage (Figure 5). **River discharge** describes the volume of water moving through a river channel over time (Figure 6). The relative contributions of surface runoff vs. groundwater seepage to river discharge depend on precipitation patterns, vegetation, topography, land use, and soil characteristics. Soon after a heavy rainstorm, river discharge increases due to surface runoff. The steady normal flow of river water is mainly from groundwater that discharges into the river. Gravity pulls river water downhill toward the ocean. Along the way the moving water of a river can erode soil particles and dissolve minerals. Groundwater also contributes a large amount of the dissolved minerals in river water. The geographic area drained by a river and its tributaries is called a **drainage basin** or **watershed**. The Mississippi River drainage basin includes approximately 40% of the U.S., a measure that includes the smaller drainage basins, such as the Ohio River and Missouri River that help to comprise it. Rivers are an important water resource for irrigation of cropland and drinking water for many cities around the world. Rivers that have had international disputes over water supply include the Colorado (Mexico, southwest U.S.), Nile (Egypt, Ethiopia, Sudan), Euphrates (Iraq, Syria, Turkey), Ganges (Bangladesh, India), and Jordan (Israel, Jordan, Syria).

Figure 5. Groundwater Seepage. Groundwater seepage can be seen in Box Canyon in Idaho, where approximately 10 cubic meters per second of seepage emanates from its vertical headwall. Source: NASA

In addition to rivers, lakes can also be an excellent source of freshwater for human use. They usually receive water from surface runoff and groundwater discharge. They tend to be short-lived on a geological time-scale because they are constantly filling in with sediment supplied by rivers. Lakes form in a variety of ways including glaciation, recent tectonic uplift (e.g., Lake Tanganyika, Africa), and volcanic eruptions (e.g., Crater Lake, Oregon). People also create artificial lakes (**reservoirs**) by damming rivers. Large changes in climate can result in major changes in a lake's size. As Earth was coming out of the last Ice Age about 15,000 years ago, the climate in the western U.S. changed from cool and moist to warm and arid, which caused more than 100 large lakes to disappear. The Great Salt Lake in Utah is a remnant of a much larger lake called Lake Bonneville.

Figure 6. River Discharge Colorado River, U.S.. Rivers are part of overland flow in the water cycle and an important surface water resource. Source: Gonzo fan2007 at Wikimedia Commons.

Although **glaciers** represent the largest reservoir of fresh water, they generally are not used as a water source because they are located too far from most people (Figure 7). Melting glaciers do provide a natural source of river water and groundwater. During the last Ice Age there was as much as %50 more water in glaciers than there is today, which caused sea level to be about 100 m lower. Over the past century, sea level has been rising in part due to melting glaciers. If Earth's climate continues to warm, the melting glaciers will cause an additional rise in sea level.

Figure 7. Mountain Glacier in Argentina Glaciers are the largest reservoir of fresh water but they are not used much as a water resource directly by society because of their distance from most people. Source: Luca Galuzzi – www.galuzzi.it

Groundwater Resources: Larger Reservoir of Usable Fresh Water

Although most people in the world use surface water, groundwater is a much larger reservoir of usable fresh water, containing more than 30 times more water than rivers and lakes combined. Groundwater is a particularly important resource in arid climates, where surface water may be scarce. In addition, groundwater is the primary water source for rural homeowners, providing 98% of that water demand in the U.S.. **Groundwater** is water located in small spaces, called **pore space**, between mineral grains and fractures in subsurface earth materials (rock or sediment). Most groundwater originates from rain or snowmelt, which infiltrates the ground and moves downward until it reaches the **saturated zone** (where groundwater completely fills pore spaces in earth materials).

Other sources of groundwater include seepage from surface water (lakes, rivers, reservoirs, and swamps), surface water deliberately pumped into the ground, irrigation, and underground wastewater treatment systems (septic tanks). **Recharge areas** are locations where surface water infiltrates the ground rather than running into rivers or evaporating. Wetlands, for example, are excellent recharge areas. A large area of sub-surface, porous rock that holds water is an aquifer. Aquifers are commonly drilled, and wells installed, to provide water for agriculture and personal use.

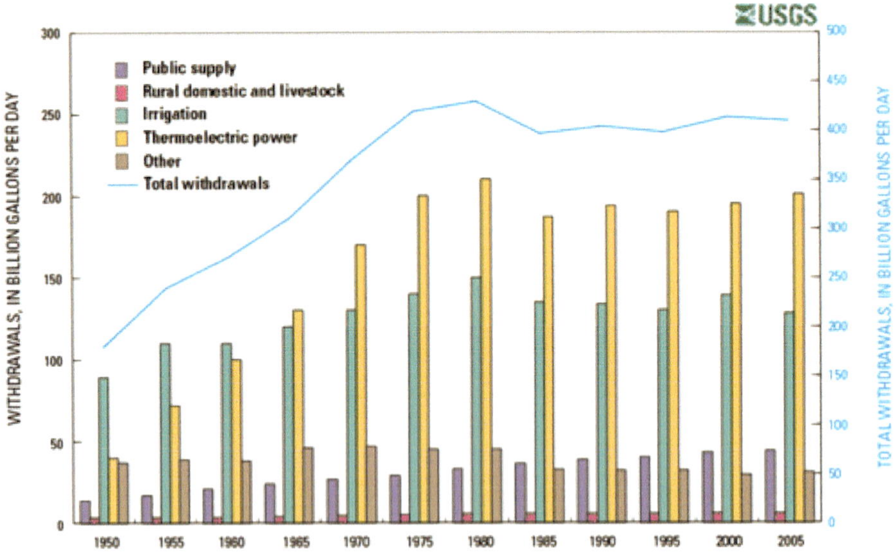

Figure 8. Trends in Total Water Withdrawals by Water-use Category, 1950-2005 Trends in total water withdrawals in the U.S. from 1950 to 2005 by water use category, including bars for thermoelectric power, irrigation, public water supply, and rural domestic and livestock. Thin blue line represents total water withdrawals using vertical scale on right. Source: United States Geological Survey

Worldwide and U.S. Water Use

People need water, oftentimes large quantities, to produce the food, energy, and mineral resources they use. Consider, for example, these approximate water requirements for some things people in the developed world use every day: one tomato = 3 gallons; one kilowatt-hour of electricity from a thermoelectric power plant = 21 gallons; one loaf of bread = 150 gallons; one pound of beef = 1,600 gallons; and one ton of steel = 63,000 gallons. Human beings require only about 1 gallon per day to survive, but a typical person in a U.S. household uses approximately 100 gallons per day, which includes cooking, washing dishes and clothes, flushing the toilet, and bathing. The water demand of an area is a function of the population and other uses of water.

Global total water use is steadily increasing at a rate greater than world population growth (Figure 10). During the 20th century global population tripled and water demand grew by a factor of six. The increase in global water demand beyond the rate of population growth is due to improved standard of living without an offset by water conservation. Increased production of goods and energy entails a large increase in water demand. The major global water uses are irrigation (68%), public supply (21%), and industry (11%).

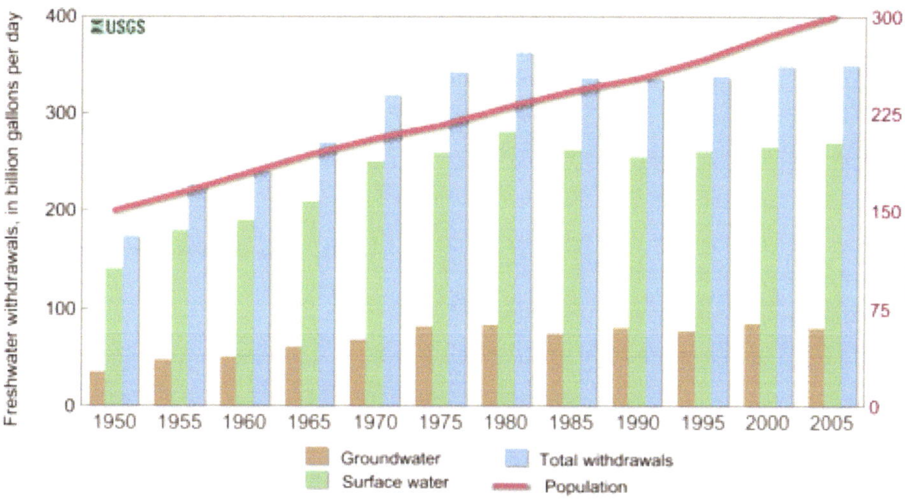

Figure 9. Trends in Source of Fresh Water Withdrawals in the U.S. from 1950 to 2005 Trends in source of fresh water withdrawals in the U.S. from 1950 to 2005, including bars for surface water, groundwater, and total water. Red line gives U.S. population using vertical scale on right. Source: United States Geological Survey

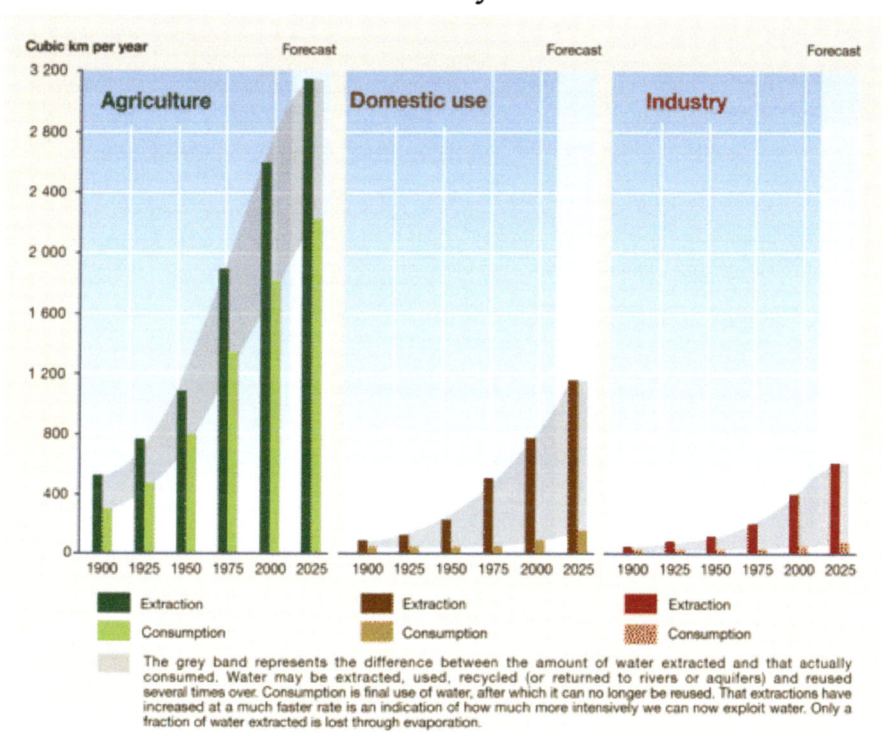

Figure 10. Trends in World Water Use from 1900 to 2000 and Projected to 2025 For each water major use category, including trends for agriculture, domestic

use, and industry. Darker colored bar represents total water extracted for that use category and lighter colored bar represents water consumed (i.e., water that is not quickly returned to surface water or groundwater system) for that use category. Source: Igor A. Shiklomanow, State Hydrological Institute (SHI, St. Petersburg) and United Nations Educational, Scientific and Cultural Organisation (UNESCO, Paris), 1999

7.2 ISSUE WITH WATER SUPPLY AND SOLUTIONS

Water Supply Problems: Resource Depletion

As groundwater is pumped from water wells, there usually is a localized drop in the water table around the well called a cone of depression. When there are a large number of wells that have been pumping water for a long time, the regional water table can drop significantly. This is called **groundwater mining**, which can force the drilling of deeper, more expensive wells that commonly encounter more saline groundwater. Rivers, lakes, and artificial lakes (reservoirs) can also be depleted due to overuse. Some large rivers, such as the Colorado in the U.S. and Yellow in China, run dry in some years. The case history of the Aral Sea discussed later in this chapter involves depletion of a lake. Finally, glaciers are being depleted due to accelerated melting associated with global warming over the past century.

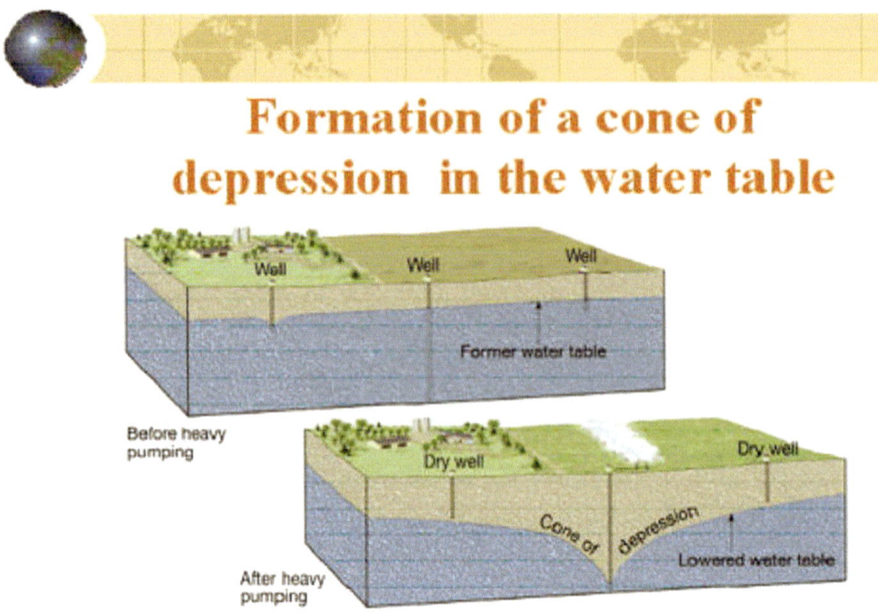

Figure 11. Formation of a Cone of Depression around a Pumping Water Well Source: Fayette County Groundwater Conservation District, TX

Another water resource problem associated with groundwater mining is saltwater intrusion, where overpumping of fresh water aquifers near ocean coastlines causes saltwater to enter fresh water zones. The drop of the water table around a **cone of depression** in an unconfined aquifer can change the direction of regional groundwater flow, which could send nearby pollution toward the pumping well instead of away from it. Finally, problems of **subsidence** (gradual sinking of the land surface over a large area) and **sinkholes** (rapid sinking of the land surface over a small area) can develop due to a drop in the water table.

Crisis in Water Supply

The **water crisis** refers to a global situation where people in many areas lack access to sufficient water, clean water, or both. This section describes the global situation involving water shortages, also called **water stress.** In general, water stress is greatest in areas with very low precipitation (major deserts), large population density (e.g., India), or both. Future global warming could worsen the water crisis by shifting precipitation patterns away from humid areas and by melting mountain glaciers that recharge rivers downstream. Melting glaciers will also contribute to rising sea level, which will worsen saltwater intrusion in aquifers near ocean coastlines.

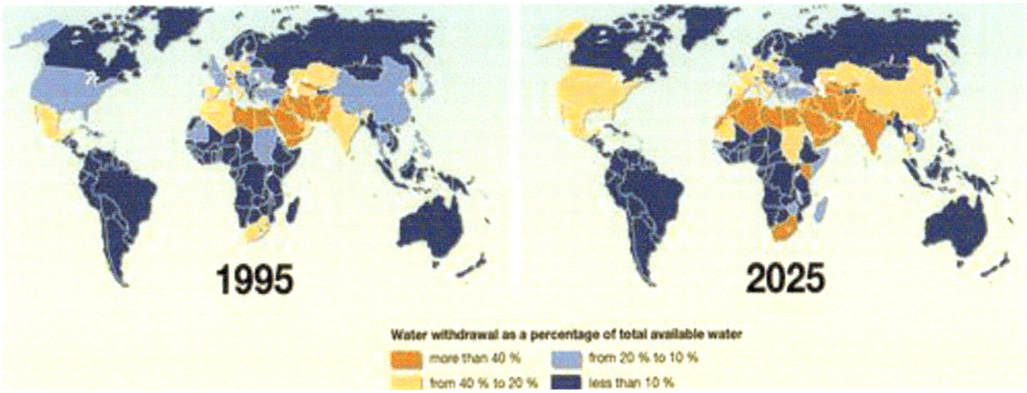

Figure 12. Countries Facing Water Stress in 1995 and Projected in 2025 Water stress is defined as having a high percentage of water withdrawal compared to total available water in the area. Source: Philippe Rekacewicz (Le Monde diplomatique), February 2006

According to a 2006 report by the United Nations Development Programme, 700 million people (11% of the world's population) lived with water stress. Most of them live in the Middle East and North Africa. By 2025, the report projects that more than 3 billion people (about 40% of the world's population) will live in water-stressed areas with the large increase coming mainly from China and India. The water crisis will also impact food production and our ability to feed the ever-growing population. We can expect future global tension and even conflict associated with water shortages and pollution. Historic and future areas of water conflict include the Middle East (Euphrates and Tigris River conflict among Turkey, Syria, and Iraq; Jordan River conflict among Israel, Lebanon, Jordan, and the Palestinian territories), Africa (Nile River conflict among Egypt, Ethiopia, and Sudan), Central Asia (Aral Sea conflict among Kazakhstan, Uzbekistan, Turkmenistan, Tajikistan, and Kyrgyzstan), and south Asia (Ganges River conflict between India and Pakistan).

The Water Supply Crisis: Sustainable Solutions?

The current and future water crisis described above requires multiple approaches to extending our fresh water supply and moving towards sustainability. Some of the longstanding traditional approaches include dams and aqueducts.

Reservoirs that form behind dams in rivers can collect water during wet times and store it for use during dry spells. They also can be used for urban water supplies. Other benefits of dams and reservoirs are hydroelectricity, flood control, and recreation. Some of the drawbacks are evaporative loss of water in arid climates, downstream river channel erosion, and impact on the ecosystem including a change from a river to lake habitat and interference with migration and spawning of fish.

Figure 13. Hoover Dam, Nevada, U.S. Hoover Dam, Nevada, U.S.. Behind the dam is Lake Mead, the largest reservoir in U.S.. White band reflects the lowered water levels in the reservoir due to drought conditions from 2000 – 2010. Source: Cygnusloop99 at Wikimedia Commons

Aqueducts can move water from where it is plentiful to where it is needed. Aqueducts can be controversial and politically difficult especially if the water transfer distances are large. One drawback is the water diversion can cause drought in the area from where the water is drawn. For example, Owens Lake and Mono Lake in central California began to disappear after their river flow was diverted to the Los Angeles aqueduct. Owens Lake remains almost completely dry, but Mono Lake has recovered more significantly due to legal intervention.

One method that can actually increase the amount of fresh water on Earth is **desalination**, which involves removing dissolved salt from seawater or saline groundwater. There are several ways to desalinate seawater including boiling, filtration, and electrodialysis. All of these procedures are moderately to very expensive and require considerable energy input, making the water produced much more expensive than fresh water from conventional sources. In addition, the process creates highly saline wastewater, which must be disposed of and creates significant environmental impact. Desalination is most common in the Middle East, where energy from oil is abundant but water is scarce.

Figure 14. The California Aqueduct California Aqueduct in southern California, U.S. Source: David Jordan at en.wikipedia

Conservation means using less water and using it more efficiently. Around the home, conservation can involve both engineered features, such as high-efficiency clothes washers and low-flow showers and toilets, as well as behavioral decisions, such as growing native vegetation that require little irrigation in desert climates, turning off the water while you brush your teeth, and fixing leaky faucets.

Rainwater harvesting involves catching and storing rainwater for reuse before it reaches the ground. Another important technique is **efficient irrigation,** which is extremely important because irrigation accounts for a much larger water demand than public water supply. Water conservation strategies in agriculture include growing crops in areas where the natural rainfall can support them, more efficient irrigation systems such as drip systems that minimize losses due to evaporation, no-till farming that reduces evaporative losses by covering the soil, and reusing treated wastewater from sewage treatment plants. Recycled wastewater has also been used to recharge aquifers.

7.3 WATER POLLUTION AND HUMAN HEALTH

The global water crisis also involves water pollution. For water to be useful for drinking and irrigation, it must not be polluted beyond certain thresholds. According to the World Health Organization, in 2008 approximately 880 million people in the world (or 13% of world population) did not have access to safe drinking water. At the same time, about 2.6 billion people (or 40% of world population) lived without improved sanitation, which is defined as having access to a public sewage system, septic tank, or even a simple pit latrine. Each year approximately 1.7 million people die from diarrheal diseases associated with unsafe drinking water, inadequate sanitation, and poor hygiene. Almost all of these deaths are in developing countries, and around 90% of them occur among children under the age of 5 (Figure 1). Compounding the water crisis is the issue of social justice; poor people more commonly lack clean water and sanitation than wealthy people in similar areas. Globally, improving water safety, sanitation, and hygiene could prevent up to 9% of all disease and 6% of all deaths.

In addition to the global waterborne disease crisis, chemical pollution from agriculture, industry, cities, and mining threatens global water quality. Some chemical pollutants have serious and well-known health effects, whereas many others have poorly known long-term health effects. In the U.S. currently more than 40,000 water bodies fit the definition of "impaired" set by EPA, which means they could neither support a healthy ecosystem nor meet water quality standards. In Gallup public polls conducted over the past decade Americans consistently put water pollution and water supply as the top environmental concerns over issues such as air pollution, deforestation, species extinction, and global warming.

Any natural water contains dissolved chemicals, some of which are important human nutrients while others can be harmful to human health. The concentration of a water pollutant is commonly given in very small units such as parts per million (ppm) or even parts per billion (ppb). An arsenic concentration of 1 ppm means 1 part of arsenic per million parts of water. This is equivalent to one drop of arsenic in 50 liters of water. To give you a different perspective on appreciating small concentration units, converting 1 ppm to length units is 1 cm (0.4 in) in 10 km (6 miles) and converting 1 ppm to time units is 30 seconds in a year. Total dissolved solids (TDS) represent the total amount of dissolved material in water. Average TDS values for rainwater, river water, and seawater are about 4 ppm, 120 ppm, and 35,000 ppm, respectively.

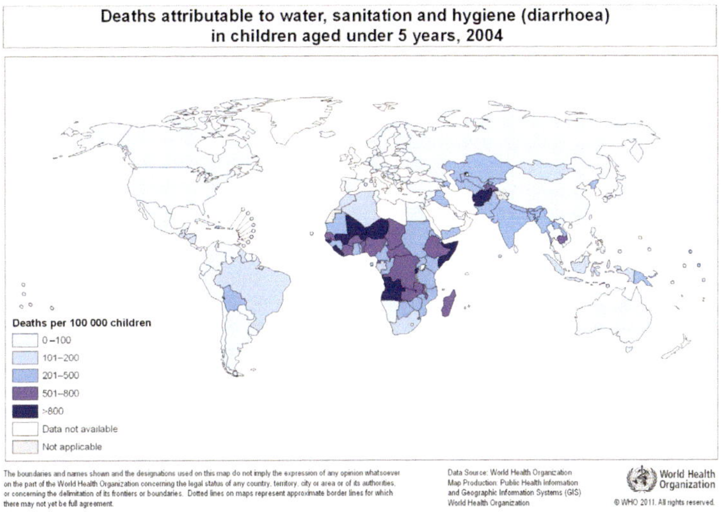

Figure 15. This work was created by the World Health Organization.

Water Pollution: An Overview

Figure 16. Water Pollution. Obvious water pollution in the form of floating debris; invisible water pollutants sometimes can be much more harmful than visible ones. Source: Stephen Codrington at Wikimedia Commons

Water pollution is the contamination of water by an excess amount of a substance that can cause harm to human beings and/or the ecosystem. The level of water pollution depends on the abundance of the pollutant, the ecological impact of the pollutant, and the use of the water. Pollutants are derived from biological, chemical, or physical processes. Although natural processes such as volcanic eruptions or evaporation sometimes can cause water pollution, most pollution is derived from human, land-based activities (Figure 2). Water pollutants can move through different water reservoirs, as the water carrying them progresses through stages of the water cycle (Figure 3). **Water residence time** (the average time that a water molecule spends in a water reservoir) is very important to pollution problems because it affects pollution potential. Water in rivers has a relatively short residence time, so pollution usually is there only briefly. Of course, pollution in rivers may simply move to another reservoir, such as the ocean, where it can cause further problems. Groundwater is typically characterized by slow flow and longer residence time, which can make groundwater pollution particularly problematic. Finally, **pollution residence time** can be much greater than the water residence time because a pollutant may be taken up for a long time within the ecosystem or absorbed onto sediment.

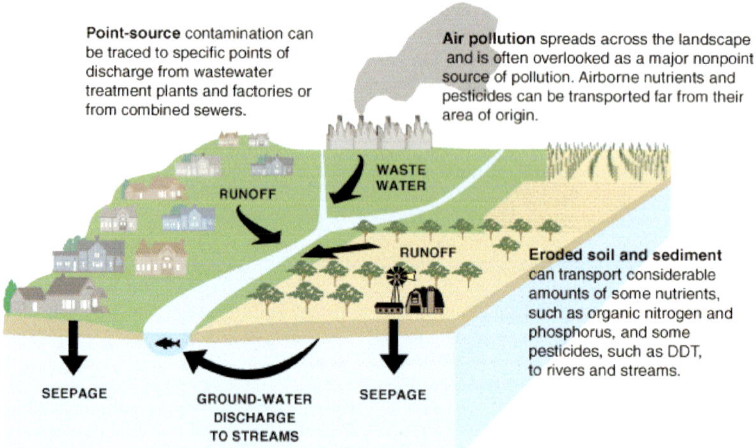

Figure 17. Sources of Water Contamination. Sources of some water pollutants and movement of pollutants into different water reservoirs of the water cycle. Source: U.S. Geological Survey

Pollutants enter water supplies from point sources, which are readily identifiable and relatively small locations, or nonpoint sources, which are large and more diffuse areas. Point sources of pollution include animal factory farms (Figure 4) that raise a large number and high density of livestock such as cows, pigs, and chickens. Also, pipes included are pipes from a factories or sewage treatment plants. Combined sewer systems that have a single set of underground pipes to collect both sewage and storm water runoff from streets for wastewater treatment can be major point sources of pollutants. During heavy rain, storm water runoff may exceed sewer capacity, causing it to back up and spilling untreated sewage directly into surface waters (Figure 5).

Figure 18. Large animal farms are often referred to as concentrated feeding operations (CFOs). These farms are considered potential point sources of pollution because untreated animal waste may enter nearby waterbodies as untreated sewage. Credit: ehp.gov

Nonpoint sources of pollution include agricultural fields, cities, and abandoned mines. Rainfall runs over the land and through the ground, picking up pollutants such as herbicides, pesticides, and fertilizer from agricultural fields and lawns; oil, antifreeze, animal waste, and road salt from urban areas; and acid and toxic elements from abandoned mines. Then, this pollution is carried into surface water bodies and groundwater. Nonpoint source pollution, which is the leading cause of water pollution in the U.S., is usually much more difficult and expensive to control than point source pollution because of its low concentration, multiple sources, and much greater volume of water.

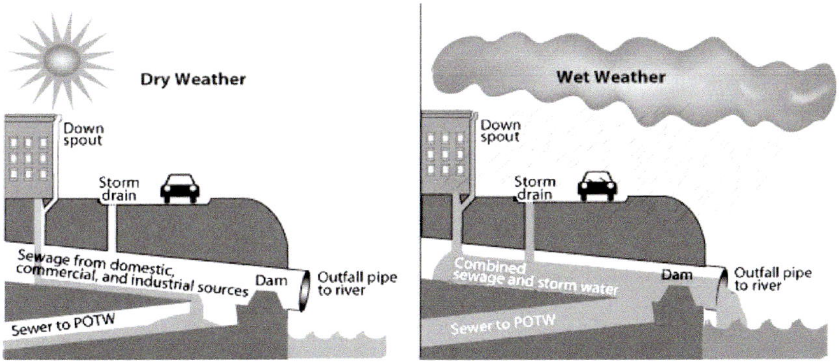

Figure 19. Combined Sewer System A combined sewer system is a possible major point source of water pollution during heavy rain due to overflow of untreated sewage. During dry weather (and small storms), all flows are handled by the

publicly owned treatment works (POTW). During large storms, the relief structure allows some of the combined stormwater and sewage to be discharged untreated to an adjacent water body. Source: U.S. Environmental Protection Agency at Wikimedia Commons

Different Types of Water Pollutants

Oxygen-demanding waste is an extremely important pollutant to ecosystems. Most surface water in contact with the atmosphere has a small amount of dissolved oxygen, which is needed by aquatic organisms for cellular respiration. Bacteria decompose dead organic matter and remove dissolved oxygen (O_2) according to the following reaction:

$$\text{organic matter} + O_2 \rightarrow CO_2 + H_2O$$

Too much decaying organic matter in water is a pollutant because it removes oxygen from water, which can kill fish, shellfish, and aquatic insects. The amount of oxygen used by aerobic (in the presence of oxygen) bacterial decomposition of organic matter is called biochemical oxygen demand (BOD). The major source of dead organic matter in many natural waters is sewage; grass and leaves are smaller sources. An unpolluted water body with respect to BOD is a turbulent river that flows through a natural forest. Turbulence continually brings water in contact with the atmosphere where the O2 content is restored. The dissolved oxygen content in such a river ranges from 10 to 14 ppm O2, BOD is low, and clean-water fish such as trout. A polluted water body with respect to oxygen is a stagnant deep lake in an urban setting with a combined sewer system. This system favors a high input of dead organic carbon from sewage overflows and limited chance for water circulation and contact with the atmosphere. In such a lake, the dissolved O2 content is ≤5 ppm O2, BOD is high, and low O2-tolerant fish, such as carp and catfish dominate.

Excessive plant nutrients, particularly nitrogen (N) and phosphorous (P), are pollutants closely related to oxygen-demanding waste. Aquatic plants require about 15 nutrients for growth, most of which are plentiful in water. N and P are called limiting nutrients, however, because they usually are present in water at low concentrations and therefore restrict the total amount of plant growth. This explains why N and P are major ingredients in most fertilizer. High concentrations of N and P from human sources (mostly agricultural and urban runoff including fertilizer, sewage, and phosphorus-based detergent) can cause cultural eutrophication, which leads to the rapid growth of aquatic producers, particularly algae. Thick mats of floating algae or rooted plants lead to a form of water pollution that damages the ecosystem by clogging fish gills and blocking sunlight. A small percentage of algal species produce toxins that can kill animals, including humans. Exponential growths of these algae are called harmful algal blooms. When the prolific algal layer dies, it becomes oxygen-demanding waste, which can create very low O2 concentrations in the water (< 2 ppm O2), a condition called hypoxia. This results in a dead zone because it causes death from asphyxiation to organisms that are unable to leave that environment. An estimated 50% of lakes in North America, Europe, and Asia are negatively impacted by cultural eutrophication. In addition, the size and number of marine hypoxic zones have grown dramatically over the past 50 years including a very large dead zone located offshore Louisiana in the Gulf of Mexico. Cultural eutrophication and hypoxia are difficult to combat, because they are caused primarily by nonpoint source pollution, which is difficult to regulate, and N and P, which are difficult to remove from wastewater.

Pathogens are disease-causing microorganisms, e.g., viruses, bacteria, parasitic worms, and protozoa, which cause a variety of intestinal diseases such as dysentery, typhoid fever, and cholera. Pathogens are the major cause of the water pollution crisis discussed at the beginning of this section. Unfortunately nearly a billion people around the world are exposed to waterborne pathogen pollution daily and around 1.5 million children mainly in underdeveloped countries die every year of waterborne diseases from pathogens. Pathogens enter water primarily from human and animal fecal waste due to inadequate sewage treatment. In many underdeveloped countries, sewage is discharged into local waters either untreated or after only rudimentary treatment. In developed countries untreated sewage discharge can occur from overflows of combined sewer systems, poorly managed livestock factory farms, and leaky or broken sewage collection systems. Water with pathogens can be remediated by adding chlorine or ozone, by boiling, or by treating the sewage in the first place.

Oil spills are another kind of organic pollution. Oil spills can result from supertanker accidents such as the Exxon Valdez in 1989, which spilled 10 million gallons of oil into the rich ecosystem of coastal Alaska and killed massive numbers of animals. The largest marine oil spill was the Deepwater Horizon disaster, which began with a natural gas explosion (Figure 6) at an oil well 65 km offshore of Louisiana and flowed for 3 months in 2010, releasing an estimated 200 million gallons of oil. The worst oil spill ever occurred during the Persian Gulf war of 1991, when Iraq deliberately dumped approximately 200 million gallons of oil in offshore Kuwait and set more than 700 oil well fires that released enormous clouds of smoke and acid rain for over nine months.

Figure 20. Deepwater Horizon Explosion Boats fighting the fire from an explosion at the Deepwater Horizon drilling rig in Gulf of Mexico offshore Louisiana on April 20, 2010. Source: United States Coast Guard via Wikimedia Commons

During an oil spill on water, oil floats to the surface because it is less dense than water, and the lightest hydrocarbons evaporate, decreasing the size of the spill but polluting the air. Then, bacteria begin to decompose the remaining oil, in a process that can take many years. After several months only about 15% of the original volume may remain, but it is in thick asphalt lumps, a form that is particularly harmful to birds, fish, and shellfish. Cleanup operations can include skimmer ships that vacuum oil from the water surface (effective only for small spills), controlled burning (works only in early stages before the light, ignitable part evaporates but also pollutes the air), **dispersants** (detergents that break up oil to accelerate its decomposition, but some dispersants may be toxic to the ecosystem), and bioremediation (adding microorganisms that specialize in quickly decomposing oil, but this can disrupt the natural ecosystem).

Toxic chemicals involve many different kinds and sources, primarily from industry and mining. General kinds of toxic chemicals include hazardous chemicals and persistent organic pollutants that include DDT (pesticide), dioxin (herbicide by-product), and PCBs (polychlorinated biphenyls, which were used as a liquid insulator in electric transformers). **Persistent organic pollutants** (POPs) are long-lived in the environment, biomagnify through the food chain, and can be toxic. Another category of toxic chemicals includes radioactive materials such as cesium, iodine, uranium, and radon gas, which can result in long-term exposure to radioactivity if it gets into the body. A final group of toxic chemicals is heavy metals such as lead, mercury, arsenic, cadmium, and chromium, which can accumulate through the food chain. Heavy metals are commonly produced by industry and at metallic ore mines. Arsenic and mercury are discussed in more detail below.

Arsenic (As) has been famous as an agent of death for many centuries. Only recently have scientists recognized that health problems can be caused by drinking small arsenic concentrations in water over a long time. It enters the water supply naturally from weathering of arsenic-rich minerals and from human activities such as coal burning and smelting of metallic ores. The worst case of arsenic poisoning occurred in the densely populated impoverished country of Bangladesh, which had experienced 100,000s of deaths from diarrhea and cholera each year from drinking surface water contaminated with pathogens due to improper sewage treatment. In the 1970s the United Nations provided aid for millions of shallow water wells, which resulted in a dramatic drop in pathogenic diseases. Unfortunately, many of the wells produced water naturally rich in arsenic. Tragically, there are an estimated 77 million people (about half of the population) who inadvertently may have been exposed to toxic levels of arsenic in Bangladesh as a result. The World Health Organization has called it the largest mass poisoning of a population in history.

Mercury (Hg) is used in a variety of electrical products, such as dry cell batteries, fluorescent light bulbs, and switches, as well as in the manufacture of paint, paper, vinyl chloride, and fungicides. Mercury acts on the central nervous system and can cause loss of sight, feeling, and hearing as well as nervousness, shakiness, and death. Like arsenic, mercury enters the water supply naturally from weathering of mercury-rich minerals and from human activities such as coal burning and metal processing. A famous mercury poisoning case in Minamata, Japan involved methylmercury-rich industrial discharge that caused high Hg levels in fish. People in the local fishing villages ate fish up to three times per day for over 30 years, which resulted in over 2,000 deaths. During that time the responsible company and national government did little to mitigate, help alleviate, or even acknowledge the problem.

Hard water contains abundant calcium and magnesium, which reduces its ability to develop soapsuds and enhances scale (calcium and magnesium carbonate minerals) formation on hot water equipment. Water softeners remove calcium and magnesium, which allows the water to lather easily and resist scale formation. Hard water develops naturally from the dissolution of calcium and magnesium carbonate minerals in soil; it does not have negative health effects in people.

Groundwater pollution can occur from underground sources and all of the pollution sources that contaminate surface waters. Common sources of groundwater pollution are leaking underground storage tanks for fuel, septic tanks, agricultural activity, landfills, and fossil fuel extraction. Common groundwater pollutants include nitrate, pesticides, volatile organic compounds, and petroleum products. Another troublesome feature of groundwater pollution is that small amounts of certain pollutants, e.g., petroleum products and organic solvents, can contaminate large areas. In Denver, Colorado 80 liters of several organic solvents contaminated 4.5 trillion liters of groundwater and produced a 5 km long contaminant plume. A major threat to groundwater quality is from underground fuel storage tanks. Fuel tanks commonly are stored underground at gas stations to reduce explosion hazards. Before 1988 in the U.S. these storage tanks could be made of metal, which can corrode, leak, and quickly contaminate local groundwater. Now, leak detectors are required and the metal storage tanks are supposed to be protected from corrosion or replaced with fiberglass tanks. Currently there are around 600,000 underground fuel storage tanks in the U.S. and over 30% still do not comply with EPA regulations regarding either release prevention or leak detection.

7.4 WATER TREATMENT PROCESS

Resolution of the global water pollution crisis requires multiple approaches to improve the quality of our fresh water and move towards sustainability. The deadliest form of water pollution, pathogenic microorganisms that cause waterborne diseases, kills almost 2 million people in underdeveloped countries every year. The best strategy for addressing this problem is proper sewage (wastewater) treatment. Untreated sewage is not only a major cause of pathogenic diseases, but also a major source of other pollutants, including oxygen-demanding waste, nutrients (N and P, particularly), and toxic heavy metals. Wastewater treatment is done at a sewage treatment plant in urban areas and through a septic tank system in rural areas.

The main purpose of **sewage (wastewater) treatment** is to remove organic matter (oxygen-demanding waste) and kill bacteria. Special methods also can be used to remove nutrients and other pollutants. The numerous steps at a conventional sewage treatment plant include **pretreatment** (screening and removal of sand and gravel), **primary treatment** (settling or floatation to remove organic solids, fat, and grease), **secondary treatment** (aerobic bacterial decomposition of organic solids), **tertiary treatment** (bacterial decomposition of nutrients and filtration), **disinfection** (treatment with chlorine, ozone, ultraviolet light, or bleach to kill most microbes), and either **discharge** to surface waters (usually a local river) or reuse for some other purpose, such as irrigation, habitat preservation, and artificial groundwater recharge (Figure 1).

The concentrated organic solid produced during primary and secondary treatment is called **sludge**, which is treated in a variety of ways including landfill disposal, incineration, use as fertilizer, and anaerobic bacterial decomposition, which is done in the absence of oxygen. Anaerobic decomposition of sludge produces methane gas, which can be used as an energy source. To reduce water pollution problems, separate sewer systems (where street runoff goes to rivers and only wastewater goes to a wastewater treatment plant) are much better than combined sewer systems, which can overflow and release untreated sewage into surface waters during heavy rain. Some cities such as Chicago, Illinois have constructed large underground caverns and also use abandoned rock quarries to hold storm sewer overflow. After the rain stops, the stored water goes to the sewage treatment plant for processing.

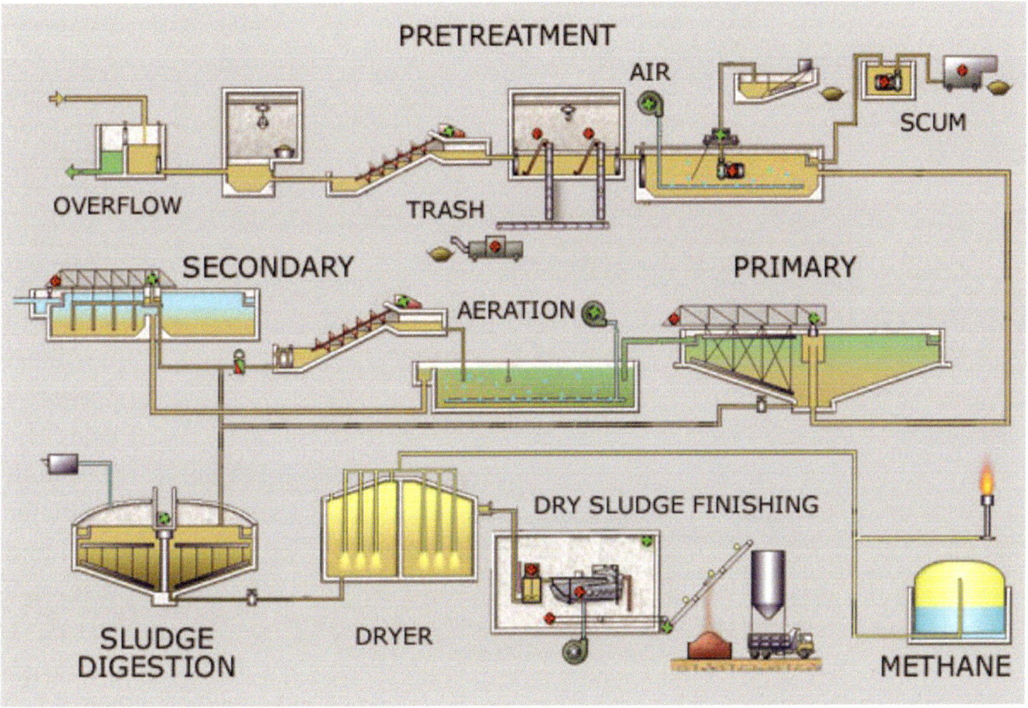

Figure 21. Steps at a Sewage Treatment Plant The numerous processing steps at a conventional sewage treatment plant include pretreatment (screening and removal of sand and gravel), primary treatment (settling or floatation to remove organic solids, fat, and grease), secondary treatment (aerobic bacterial decomposition of organic solids), tertiary treatment (bacterial decomposition of nutrients and filtration), disinfection (treatment with chlorine, ozone, ultraviolet light, or bleach), and either discharge to surface waters (usually a local river) or reuse for some other purpose, such as irrigation, habitat preservation, and artificial groundwater recharge. Source: Leonard G.via Wikipedia

A septic tank system is an individual sewage treatment system for homes in typically rural settings. The basic components of a septic tank system (Figure 2) include a sewer line from the house, a septic tank (a large container where sludge settles to the bottom and microorganisms decompose the organic solids anaerobically), and the drain field (network of perforated pipes where the clarified water seeps into the soil and is further purified by bacteria). Water pollution problems occur if the septic tank malfunctions, which usually occurs when a system is established in the wrong type of soil or maintained poorly.

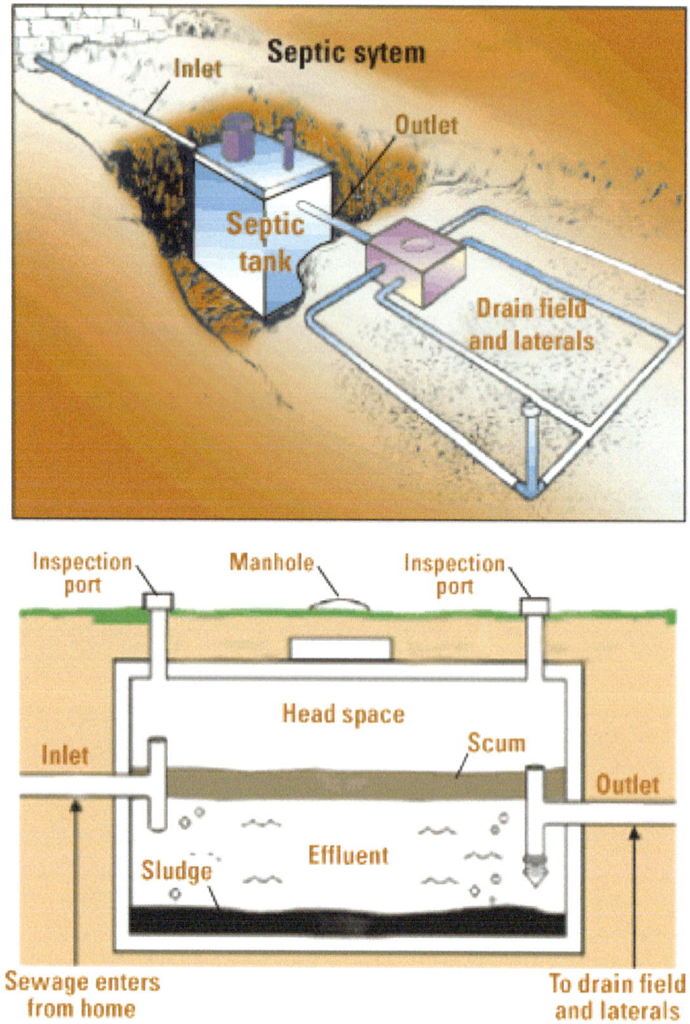

Figure 22. Septic System Septic tank system for sewage treatment. Source: United States Geological Survey

For many developing countries, financial aid is necessary to build adequate sewage treatment facilities. The World Health Organization estimates an estimated cost savings of between \$3 and \$34 for every \$1 invested in clean water delivery and sanitation. The cost savings are from health care savings, gains in work and school productivity, and prevented deaths. Simple and inexpensive techniques for treating water at home include chlorination, filters, and solar disinfection. Another alternative is to use constructed wetlands technology (marshes built to treat contaminated water), which is simpler and cheaper than a conventional sewage treatment plant.

Bottled water is not a sustainable solution to the water crisis. Bottled water is not necessarily any safer than the U.S. public water supply, it costs on average about 700 times more than U.S. tap water, and every year it uses approximately 200 billion plastic and glass bottles that have a relatively low rate of recycling. Compared to tap water, it uses much more energy, mainly in bottle manufacturing and long-distance transportation. If you don't like the taste of your tap water, then please use a water filter instead of bottled water!

CLEAN WATER ACT

Figure 23. Cuyahoga River on fire. Source: National Oceanic and Atmospheric Administration

During the early 1900s rapid industrialization in the U.S. resulted in widespread water pollution due to free discharge of waste into surface waters. The Cuyahoga River in northeast Ohio caught fire numerous times, including a famous fire in 1969 that caught the nation's attention. In 1972 Congress passed one of the most important environmental laws in U.S. history, the Federal Water Pollution Control Act, which is more commonly called the Clean Water Act. The purpose of the Clean Water Act and later amendments is to maintain and restore water quality, or in simpler terms to make our water swimmable and fishable. It became illegal to dump pollution into surface water unless there was formal permission and U.S. water quality improved significantly as a result. More progress is needed because currently the EPA considers over 40,000 U.S. water bodies as impaired, most commonly due to pathogens, metals, plant nutrients, and oxygen depletion. Another concern is protecting groundwater quality, which is not yet addressed sufficiently by federal law.

7.5 CASE STUDY: THE ARAL SEA - GOING, GOING, GONE

Fig. 24 A comparison of the Aral Sea in 1989 (left) and 2014 (right). Credit: This work is in the Public Domain, CC0

The Aral Sea is a lake located east of the Caspian Sea between Uzbekistan and Kazakhstan in central Asia. This area is part of the Turkestan desert, which is the fourth largest desert in the world; it is produced from a rain shadow effect by Afghanistan's high mountains to the south. Due to the arid and seasonally hot climate there is extensive evaporation and limited surface waters in general. Summer temperatures can reach 60°C (140°F)! The water supply to the Aral Sea is mainly from two rivers, the Amu Darya and Syr Darya, which carry snow melt from mountainous areas. In the early 1960s, the then-Soviet Union diverted the Amu Darya and Syr Darya Rivers for irrigation of one of the driest parts of Asia to produce rice, melons, cereals, and especially cotton. The Soviets wanted cotton or white gold to become a major export. They were successful, and, today Uzbekistan is one of the world's largest exporters of cotton. Unfortunately, this action essentially eliminated any river inflow to the Aral Sea and caused it to disappear almost completely.

Figure 25. Map of Aral Sea Area Map shows lake size in 1960 and political boundaries of 2011. Countries in yellow are at least partially in Aral Sea drainage basin. Source: Wikimedia Commons

In 1960, Aral Sea was the fourth largest inland water body; only the Caspian Sea, Lake Superior, and Lake Victoria were larger. Since then, it has progressively shrunk due to evaporation and lack of recharge by rivers. Before 1965, the Aral Sea received 2060 km^3 of fresh water per year from rivers and by the early 1980s it received none. By 2007, the Aral Sea shrank to about 10% of its original size and its salinity increased from about 1% dissolved salt to about 10% dissolved salt, which is 3 times more saline than seawater. These changes caused an enormous environmental impact. A once thriving fishing industry is dead as are the 24 species of fish that used to live there; the fish could not adapt to the more saline waters. The current shoreline is tens of kilometers from former fishing towns and commercial ports. Large shing boats lie in the dried up lakebed of dust and salt. A frustrating part of the river diversion project is that many of the irrigation canals were poorly built, allowing abundant water to leak or evaporate. An increasing number of dust storms blow salt, pesticides, and herbicides into nearby towns causing a variety of respiratory illnesses including tuberculosis.

Figure 26. This abandoned ship lies in a dried up lake bed that was the Aral Sea near Aral, Kazakhstan Source: Staecker at Wikimedia Commons

The wetlands of the two river deltas and their associated ecosystems have disappeared. The regional climate is drier and has greater temperature extremes due to the absence of moisture and moderating influence from the lake. In 2003, some lake restoration work began on the northern part of the Aral Sea and it provided some relief by raising water levels and reducing salinity somewhat. The southern part of the Aral Sea has seen no relief and remains nearly completely dry. The destruction of the Aral Sea is one of the planet's biggest environmental disasters and it is caused entirely by humans. Lake Chad in Africa is another example of a massive lake that has nearly disappeared for the same reasons as the Aral Sea. Aral Sea and Lake Chad are the most extreme examples of large lakes destroyed by unsustainable diversions of river water. Other lakes that have shrunk significantly due to human diversions of water include the Dead Sea in the Middle East, Lake Manchar in Pakistan, and Owens Lake and Mono Lake, both in California.

AT A GLANCE

Precipitation—a major control of fresh water availability—is unevenly distributed around the globe. More precipitation falls near the equator, and landmasses there are characterized by a tropical rainforest climate. Less precipitation tends to fall near 2030 north and south latitude, where the world's largest deserts are located. The water crisis refers to a global situation where people in many areas lack access to sufficient water or clean water or both. The current and future water crisis requires multiple approaches to extending our fresh water supply and moving towards sustainability. Some of the longstanding traditional approaches include dams and aqueducts. Water pollution is the contamination of water by an excess amount of a substance that can cause harm to human beings and the ecosystem. The level of water pollution depends on the abundance of the pollutant, the ecological impact of the pollutant, and the use of the water. The most deadly form of water pollution, pathogenic microorganisms that cause waterborne diseases, kills almost 2 million people in underdeveloped countries every year. Resolution of the global water pollution crisis requires multiple approaches to improve the quality of fresh water. The best strategy for addressing this problem is proper sewage treatment. Untreated sewage is not only a major cause of pathogenic diseases, but also a major source of other pollutants, including oxygen-demanding waste, plant nutrients, and toxic heavy metals.

CHAPTER-8

FOOD, HUNGER FOOD SECURITY

8.1 FOOD SECURITY: AN OVERVIEW

Progress continues in the fight against hunger, yet an unacceptably large number of people lack the food they need for an active and healthy life. The latest available estimates indicate that about 795 million people in the world – just over one in nine –still go to bed hungry every night, and an even greater number live in poverty (defined as living on less than $1.25 per day). Poverty—not food availability—is the major driver of food insecurity. Improvements in agricultural productivity are necessary to increase rural household incomes and access to available food but are insufficient to ensure food security. Evidence indicates that poverty reduction and food security do not necessarily move in tandem. The main problem is lack of economic (social and physical) access to food at national and household levels and inadequate nutrition (or hidden hunger). Food security not only requires an adequate supply of food but also entails availability, access, and utilization by all—people of all ages, gender, ethnicity, religion, and socioeconomic levels.

From Agriculture to Food Security

Agriculture and food security are inextricably linked. The agricultural sector in each country is dependent on the available natural resources, as well as the politics that govern those resources. **Staple food crops** are the main source of dietary energy in the human diet and include things such as rice, wheat, sweet potatoes, maize, and cassava.

Food security

Food security is essentially built on four pillars: availability, access, utilization and stability. An individual must have access to sufficient food of the right dietary mix (quality) at all times to be food secure. Those who never have sufficient quality food are chronically food insecure.

When food security is analyzed at the national level, an understanding not only of national production is important, but also of the country's **access** to food from the global market, its foreign exchange earnings, and its citizens' consumer choices. Food security analyzed at the household level is conditioned by a household's own food production and household members' ability to purchase food of the right quality and diversity in the market place. However, it is only at the individual level that the analysis can be truly accurate because only through understanding who consumes what can we appreciate the impact of sociocultural and gender inequalities on people's ability to meet their nutritional needs.The definition of food security is often applied at varying levels of aggregation, despite its articulation at the individual level. The importance of a pillar depends on the level of aggregation being addressed. At a global level, the important pillar is food **availability**. Does global agricultural activity produce sufficient food to feed all the world's inhabitants? The answer today is yes, but it may not be true in the future given the impact of a growing world population, emerging plant and animal pests and diseases, declining soil productivity and environmental quality, increasing use of land for fuel rather than food, and lack of attention to agricultural research and development, among other factors.

The third pillar, food **utilization**, essentially translates the food available to a household into nutritional security for its members. One aspect of utilization is analyzed in terms of distribution according to need. Nutritional standards exist for the actual nutritional needs of men, women, boys, and girls of different ages and life phases (that is, pregnant women), but these "needs" are often socially constructed based on culture. For example, in South Asia evidence shows that women eat after everyone else has eaten and are less likely than men in the same household to consume preferred foods such as meats and fish. **Hidden hunger** commonly results from poor food utilization: that is, a person's diet lacks the appropriate balance of macro- (calories) and micronutrients (vitamins and minerals). Individuals may look well nourished and consume sufficient calories but be deficient in key micronutrients such as vitamin A, iron, and iodine.

When food security is analyzed at the national level, an understanding not only of national production is important, but also of the country's access to food from the global market, its foreign exchange earnings, and its citizens' consumer choices. Food security analyzed at the household level is conditioned by a household's own food production and household members' ability to purchase food of the right quality and diversity in the market place. However, it is only at the individual level that the analysis can be truly accurate because only through understanding who consumes what can we appreciate the impact of sociocultural and gender inequalities on people's ability to meet their nutritional needs.

Food stability is when a population, household, or individual has access to food at all times and does not risk losing access as a consequence of cyclical events, such as the dry season. When some lacks food stability, they have malnutrition, a lack of essential nutrients. This is economically costly because it can cost individuals 10 percent of their lifetime earnings and nations 2 to 3 percent of gross domestic product (GDP) in the worst-affected countries (Alderman 2005). Achieving food security is even more challenging in the context of HIV and AIDS. HIV affects people's physical ability to produce and use food, reallocating household labor, increasing the work burden on women, and preventing widows and children from inheriting land and productive resources.

Obesity

Obesity means having too much body fat. It is not the same as overweight, which means weighing too much. Obesity has become a significant global health challenge, yet is preventable and reversible. Over the past 20 years, a global overweight/obesity epidemic has emerged, initially in industrial countries and now increasingly in low- and middle-income countries, particularly in urban settings, resulting in a triple burden of undernutrition, micronutrient deficiency, and overweight/obesity. There is significant variation by region; some have very high rates of undernourishment and low rates of obesity, while in other regions the opposite is true (Figure 1).

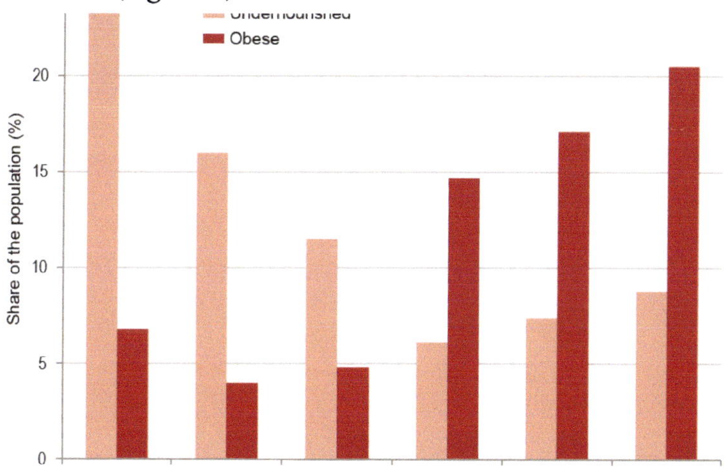

Figure 1. Obesity and undernourishment by region.

However, obesity has increased to the extent that the number of overweight people now exceeds the number of underweight people worldwide. The economic cost of obesity has been estimated at $2 trillion, accounting for about 5% of deaths worldwide. Almost 30% of the world's population, or 2.1 billion people, are overweight or obese, 62% of whom live in developing countries.

Obesity accounts for a growing level and share of worldwide noncommunicable diseases, including diabetes, heart disease, and certain cancers that can reduce quality of life and increase public health costs of already under-resourced developing countries. The number of overweight children is projected to double by 2030. Driven primarily by increasing availability of processed, affordable, and effectively marketed food, the global food system is falling short with rising obesity and related poor health outcomes. Due to established health implications and rapid increase in prevalence, obesity is now a recognized major global health challenge.

8.2 BIOTECHNOLOGY AND GENETIC ENGINEERING

Figure 2. The symptoms of papaya ringspot virus are shown on the tree (a) and fruit (b). "This work" is in the Public Domain, CC0.

In the early 1990s, an emerging disease was destroying Hawaii's production of papaya and threatening to decimate the $11 million industry (Figure 1). Fortunately, a man named Dennis Gonsalves, who was raised on a sugar plantation and then became a plant physiologist at Cornell University, would develop papaya plants genetically engineered to resist the deadly virus. By the end of the decade, the Hawaiian papaya industry and the livelihoods of many farmers were saved thanks to the free distribution of Dr. Gonsalves seeds.

The development of a new strain of crop is an example of agricultural biotechnology: a range of tools that include both traditional breeding techniques and more modern lab-based methods. Traditional methods date back thousands of years, whereas biotechnology uses the tools of genetic engineering developed over the last few decades. Genetic engineering is the name for the methods that scientists use to introduce new traits to an organism. This process results in genetically modified organisms, or GMO. For example, plants may be genetically engineered to produce characteristics to enhance the growth or nutritional profile of food crops. GMO that are crop species are commonly called genetically engineered crops, or GE crops for short

Genetic Modification of Crops: A Brief History

Nearly all the fruits and vegetables found in your local market would not occur naturally. In fact, they exist only because of human intervention that began thousands of years ago. Humans created the vast majority of crop species by using traditional breeding practices on naturally-occurring, wild plants. These practices rely upon **selective breeding** (human assisted-breeding of individuals with desirable traits). Traditional breeding practices, although low-tech and simple to perform, have the practical outcome of modifying an organism's genetic information, thus producing new traits.

An interesting example is maize (corn). Biologists have discovered that maize was developed from a wild plant called teosinte. Through traditional breeding practices, humans living thousands of years ago in what is now Southern Mexico began selecting for desirable traits until they were able to transform the plant into what is now known as maize. In doing so, they permanently (and unknowingly) altered its genetic instructions, allowing for new traits to emerge. Considering this history, we might ask the question: is there really such a thing as "non-GMO" maize?

Figure 3. A wild grass called teosinte was genetically modified through selective breeding to produce what is now known as maize (corn). This process of transformation started thousands of years ago by indigenous people of what is now Mexico. "This work" by Nicolle Rager Fuller, National Science Foundation is in the Public Domain, CC0.

Figure 4. Brassica oleracea is a plant in the mustard family and is known as wild cabbage. From it were developed many familar crops, such as cauliflower, broccoli, Brussel sprouts, and of course, cabbage."Brassica oleracea wild" by Kulac is licensed under CC BY-SA 4.0

This history of genetic modification is common to nearly all crop species. For example, cabbage, broccoli, Brussel sprouts, cauliflower, and kale were all developed from a single species of wild mustard plant (Figure 2). Wild nightshade was the source of tomatoes, eggplant, tobacco, and potatoes, the latter developed by humans 7,000 – 10,000 years ago in South America.

Traditional Breeding vs. Modern Genetic Engineering

To produce new traits in livestock, pets, crops, or other type of organism, there almost always has to be an underlying change in that organism's genetic instructions. What many people might not understand is that traditional breeding practices do, in fact, result in permanent genetic changes and is therefore a type of genetic modification. This misunderstanding may arise because traditional breeding practices do not require sophisticated laboratory equipment or any knowledge of genetics, which some may see as a perquisite for genetic modification.

How do traditional breeding practices compare to modern genetic engineering? Both result in changes to an organism's genetic information, but the magnitude of those changes varies amongst the two techniques (Figure 3). Traditional breeding shuffles all of the genes between the two organisms being bred, which can number into the tens of thousands (maize, for example, has 32,000 genes). When mixing such a large number of genes, the results can be unpredictable. Modern genetic engineering is more precise in the sense that biologists can modify just a single gene. Also, genetic engineering can introduce a gene between two distantly-related species, such as inserting a bacterial gene into a plant. Such transfer might seem unusual, but it is not without its equivalent in nature. In a process called horizontal gene transfer, DNA from one species can be inserted into a different species. One recent study, for example, has found that humans contain about 150 genes from other species, including bacteria.

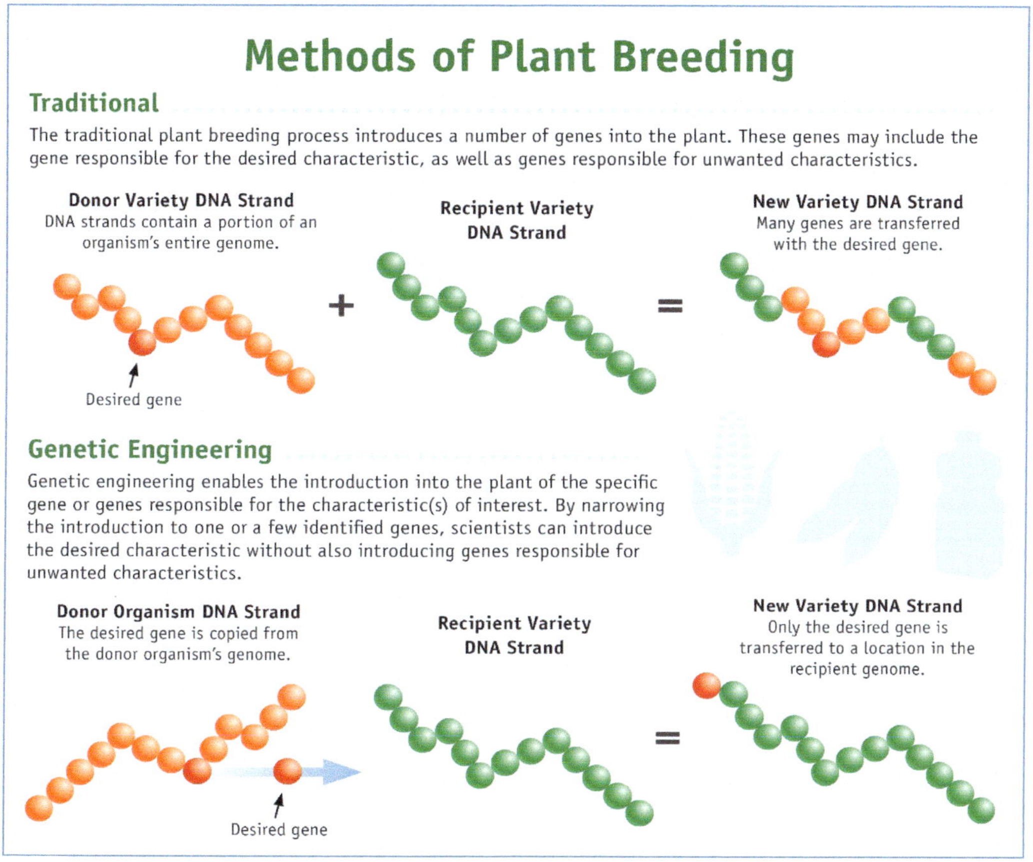

Figure 5. Both traditional breeding and modern genetic engineering produce genetic modifications. Genetic engineering allows for fewer and more precise genetic modifications. FDA graphic by Michael J. Ermarth (Methods of Plant Breeding) [CC0 Public Domain], via Wikimedia Commons.

Benefits of Genetic Engineering

Enhanced nutrition

Advances in biotechnology may provide consumers with foods that are nutritionally-enriched (Figure 4), longer-lasting, or that contain lower levels of certain naturally occurring toxins present in some food plants. For example, developers are using biotechnology to try to reduce saturated fats in cooking oils, reduce allergens in foods, and increase disease-fighting nutrients in foods. Biotechnology may also be used to conserve natural resources, enable animals to more effectively use nutrients present in feed, decrease nutrient runoff into rivers and bays, and help meet the increasing world food and land demands.

Figure 6. White rice and Golden rice. Genetically engineered "Golden Rice" contains up to 35 μg β-carotene per gram of rice, which could prevent millions of cases of blindness worldwide.

Cheaper and More Manageable Production

Biotechnology may provide farmers with tools that can make production cheaper and more manageable. For example, some biotechnology crops can be engineered to tolerate specific herbicides, which make weed control simpler and more efficient. Other crops have been engineered to be resistant to specific plant diseases and insect pests, which can make pest control more reliable and effective, and/or can decrease the use of synthetic pesticides. These crop production options can help countries keep pace with demands for food while reducing production costs.

Improved pest control and weed management

Biotechnology has helped to make both pest control and weed management safer and easier while safeguarding crops against disease. For example, genetically engineered insect-resistant cotton has allowed for a significant reduction in the use of persistent, synthetic pesticides that may contaminate groundwater and the environment. In terms of improved weed control, herbicide-tolerant soybeans, cotton, and corn enable the use of reduced-risk herbicides that break down more quickly in soil and are non-toxic to wildlife and humans.

Genetically Engineered Crops: Potential Concerns

The complexity of ecological systems presents considerable challenges for experiments that assess the risks and benefits of GE crops. Assessing such risks is difficult, because both natural and human-modified systems are highly complex and fraught with uncertainties that may not be clarified until long after an experimental introduction has been concluded. Critics of GE crops warn that their cultivation should be carefully considered within broader ecosystems because of their potential benefits and hazards to the environment.

In addition to environmental risks, some people are concerned about potential health risks of GE crops because they feel that genetic modification alters the intrinsic properties, or essence, of an organism. As discussed above, however, it is known that both traditional breeding practices and modern genetic engineering produce permanent genetic modifications. Further, traditional breeding practices actually have a larger and more unpredictable impact on a species' genetics because of its comparably crude nature. Considering this, it is wise that both new GE crops and traditionally produced crops should be studied for potential human health risks.

To address these various concerns, the US National Academies of Sciences, Engineering, and Medicine (NASEM) published a comprehensive, 500-page report in 2016 that summarized the current scientific knowledge regarding GE crops. The report, titled *Genetically Engineered Crops: Experiences and Prospects*, reviewed more than 900 research articles, in addition to public comments and expert testimony. Results from this seminal report, hereafter referred to as the **"GE Crop Report"** for brevity, is shared in the various subsections below.

Interbreeding with Native Species

Through interbreeding, or hybridization, GE crops might share their genetically-modified DNA with wild relatives. This could affect the genetics of those wild relatives and have unforeseen consequences on their populations, and could even have implications for the larger ecosystem. For example, if a gene engineered to confer herbicide resistance were to pass from a GE crop to a wild relative, it might transform the wild species into a **'super weed'** – a species that could not be controlled by herbicide. Its rampant growth could then displace other wild species and the wildlife that depends on it, thus inflecting ecological harm.

NASEM's GE Crop Report did find some evidence of genetic transfer between GE crops and wild relatives. However, there was no evidence of ecological harm from that transfer. Clearly, continued monitoring, especially for newly-developed crops, is warranted.

Consumers Right to Choose

The International Federation of Organic Agriculture Movement has made stringent efforts to keep GE crops out of organic production, yet some US organic farmers have found their corn (maize) crops, including seeds, contain detectable levels of genetically engineered DNA. The organic movement is firm in its opposition to any use of GE crops in agriculture, and organic standards explicitly prohibit their use (however, keep in mind that even "organic" maize has incurred significant genetic modification compared to its wild relative, teosinte). The farmers, whose seed is contaminated, have been under rigid organic certification, which assures that they did not use any kind of genetically modified materials on their farms.

Any trace of GE crops must have come from outside their production areas. While the exact origin is unclear at this time, it is likely that the contamination has been caused by pollen drift from GE crop fields in surrounding areas. However, the contamination may have also come from the seed supply. Seed producers, who intended to supply GE crop-free seed, have also been confronted with genetic contamination and cannot guarantee that their seed is 100% GE crop-free.

Ecological Effects: Long Term

An early study indicated the pollen from a particular type of genetically modified corn may be harmful to the caterpillars of monarch butterflies, This type of corn, known as ***Bt* corn,** is genetically modified to produce a bacterial protein that acts as an insecticide. This trait is favorable because it reduces the amount of insecticides used by farmers. Pollen from *Bt* corn can be harmful to caterpillars, but only at very high concentrations. These concentrations are seldom reached in nature and follow-up studies have found the effect of *Bt* corn to be negligible.

NASEM's GE Crop Report documents that the validity of that initial monarch study was questioned by other scientists and this ultimately led to a large, multi-national study funded by the US and Canada. They found that the vast majority of *Bt* corn grown did not represent a risk to monarchs. However, one strain of *Bt* corn did, and it was consequently removed from the market.

The GE Crop Report also mentioned a separate threat to monarch: loss of milkweed, which is critical to the butterfly's lifecycle. Some GE crops are engineered to resist the herbicide glyphosate. Farmers using these crops can spray their entire field with the herbicide, which kills milkweed but not their GE crop. This can lower the amount of milkweed growing within the habitat range of monarchs. The report concluded that more studies are needed to quantify the actual impact this may be having on monarch populations.

GE Crops and Human Health Risk

At least some of the genes used in GE crops may not have been used in the food supply before, so GM foods may pose a potential risk for human health, such as producing new allergens. But this is also true of crops generated by traditional breeding practices (because both produce genetic modifications and thus new traits).

Like other 'controversial' scientific issues, the scientific consensus on GE crops is quite clear: they are safe. The UN's Food and Agriculture Organization has concluded that risks to human and animal health from the use of GMOs are negligible. NASEM's GE Crop Report found "no substantiated evidence of a difference in risks to human health between current commercially available genetically engineered (GE) crops and conventionally bred crops, nor did it find conclusive cause-and-effect evidence of environmental problems from the GE crops." The American Medical Association's Council on Science and Public Health, in 2012, stated that "Bioengineered foods have been consumed for close to 20 years, and during that time, no overt consequences on human health have been reported and/or substantiated in the peer-reviewed literature." Similar statements have been made by the US National Resource Council and the American Association for the Advancement of Science, which publishes the preeminent scholarly journal, *Science*.

The potential of GE crops to be allergenic is one of the potential adverse health effects and it should continue to be studied, especially because some scientific evidences indicates that animals fed GE crops have been harmed. NASEM's GE Crop Report concluded that when developing new crops, it is the product that should be studied for potential health and environmental risks, not the process that achieved that product. What this means is, because both traditional breeding practices and modern genetic engineering produce new traits through genetic modification, they both present potential risks. Thus, for the safety of the environment and human health, both should be adequately studied.

Intellectual Property Rights and GE Crops

Intellectual property rights are one of the important factors in the current debate on GE crops. GE crops can be patented by Agri-businesses, which can lead to them controlling and potentially exploiting agricultural markets. Some accuse companies, such as Monsanto, of allegedly controlling seed production and pricing, much to the detriment of farmers. NASEM's GE Crop Report recommends more research into how the concentration of seed-markets by a few companies, and the subsequent reduction of free market competition, may be affecting seed prices and farmers.

It should be noted that crops developed by traditional breeding can also be legally protected and controlled in ways similar to GE crops. Jim Myers, from Oregon State University notes that "In all but a few cases, all contemporary varieties developed by private breeders are [legally] protected, and most public varieties are protected as well."

Are GE Crops the Solution We Need?

Significant resources, both financial and intellectual, have been allocated to answering the question: are GE crops safe? After many hundreds of scientific studies, the answer is yes. But a significant question still remains: are they necessary? Certainly, such as in instances like Hawaii's papaya, which were threated with eradication due to an aggressive disease, genetic engineering was a quick and effective solution that would have been extremely difficult, if not impossible, to solve using traditional breeding practices.

However, in many cases, the early promises of GE crops – that they would improve nutritional quality of foods, confer disease resistance, and provide unparalleled advances in crop yields – have largely failed to come to fruition. NASEM's GE Crop Report states that while GE crops have resulted in the reduction of agricultural loss from pests, reduced pesticide use, and reduced rates of injury from insecticides for farm workers, they have not increased the rate at which crop yields are advancing when compared to non-GE crops. Additionally, while there are some notable exceptions like golden rice or virus-resistant papayas, very few GE crops have been produced to increase nutritional capacity or to prevent plant disease that can devastate a farmer's income and reduce food security. The vast majority of GE crops are developed for only two purposes: to introduce herbicide resistance or pest resistance.

Genetic engineering of crops represents in important tool in a world of rapidly changing climate and a burgeoning human population, but as you will see in the next chapter, it is only one of many tools that agriculturists can use to produce enough food for all humans while simultaneously working to conserve the environment.

AT A GLANCE

Progress continues in the fight against hunger, yet an unacceptably large number of people still lack the food they need for an active and healthy life. About 795 million people in the world still go to bed hungry every night, and an even greater number live in poverty. Poverty is the major driver of food insecurity. Improvements in agricultural productivity are necessary to increase rural household incomes and access to available food but are insufficient to ensure food security. Food security is essentially built on four pillars: availability, access, utilization and stability. Women are crucial in the translation of the products of a vibrant agriculture sector into food and nutritional security for their households. They are often the farmers who cultivate food crops and produce commercial crops alongside the men in their households as a source of income. Over the past 20 years, a global obesity epidemic has emerged. Due to established health implications and rapid increase in prevalence, obesity is now a recognized major global health challenge, and no national success stories in curbing its growth have so far been reported. Genetic engineering is the name for methods that scientists use to introduce new traits or characteristics to an organism. Advocates say that application of genetic engineering in agriculture has resulted in benefits to farmers, producers, and consumers. Critics advise that the risks for the introduction of a GMO into each new ecosystem need to be examined on a case-by-case basis, alongside appropriate risk management measures.

CHAPTER-9

CONVENTIONAL FARMING SYSTEMS AND SUSTAINABLE AGRICULTURE

9.1 SOIL PROFILES & FPORMING PROCESSES

Soil

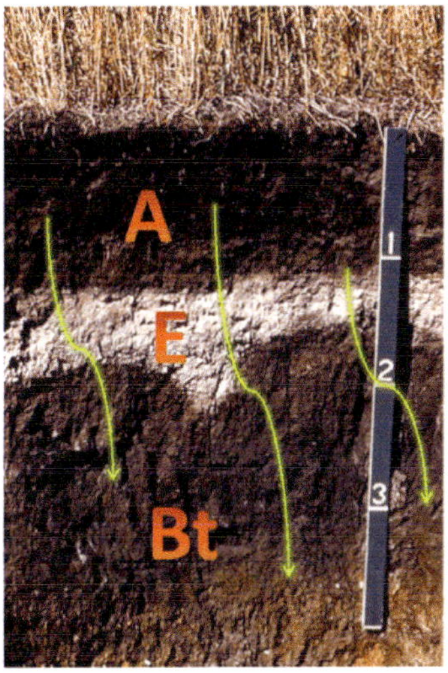

Figure 1. Soil Profile. Photograph shows a soil profile from South Dakota with A, E, and Bt horizons. The yellow arrows symbolize translocation of fine clays to the Bt horizon. The scale is in feet. Source: University of Idaho and modified by D. Grimley.

The word "**soil**" has been defined differently by different scientific disciplines. In agriculture and horticulture, soil generally refers to the medium for plant growth, typically material within the upper meter or two (Figure 1). We will use this definition in this chapter. Soil consists predominantly of mineral matter, but also contains organic matter (**humus**) and living organisms. The pore spaces between mineral grains are filled with varying proportions of water and air.

In common usage, the term soil is sometimes restricted to only the dark topsoil in which we plant our seeds or vegetables. In a more broad definition, civil engineers use the term soil for any unconsolidated (soft when wet) material that is not considered bedrock. Under this definition, soil can be as much as several hundred feet thick! Ancient soils, sometimes buried and preserved in the subsurface, are referred to as **paleosols** (Figure 2) and reflect past climatic and environmental conditions.

Importance of Soil

Figure 2. Modern versus Buried Soil Profiles. A buried soil profile, or paleosol (above geologist 's head), represents soil development during the last interglacial period. A modern soil profile (Alfisol) occurs near the land surface. Source: D. Grimley.

Soil is important to our society primarily because it provides the foundation of agriculture and forestry. Of course, soil is also a critical component for terrestrial ecosystems, and thus important to animals, plants, fungi, and microorganisms.

Soil plays a role in nearly all biogeochemical cycles on the Earth's surface. Global cycling of key elements such as carbon (C), nitrogen (N), sulfur (S), and phosphorous (P) all pass through soil. In the hydrologic cycle, soil helps to mediate the flow of precipitation from the surface into the groundwater. Microorganisms living in soil can also be important components of biogeochemical cycles through the action of decomposition and other processes such as nitrogen fixation.

Soil Forming Factors

The fundamental factors that affect soil genesis can be categorized into five elements: climate, organisms, relief, parent material, and time. One could say that the relief, climate, and organisms dictate the local soil environment and act together to cause weathering and mixing of the soil parent material over time. As soil is formed it often has distinct layers, which are formally described as "horizons." Upper horizons (labeled as the A and O horizons) are richer in organic material and so are important in plant growth, while deeper layers (such as the B and C horizons) retain more of the original features of the bedrock below (Figure 3).

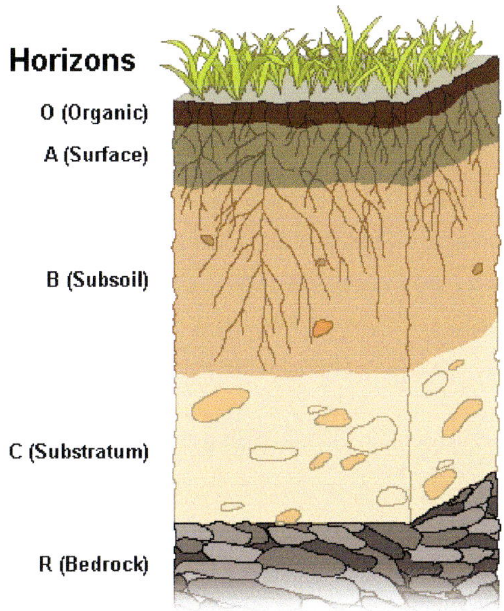

Figure 3. This images shows the different horizons, or layers, in soil. This work by Wilsonbiggs is licensed under CC BY-SA 4.0

Climate

The role of climate in soil development includes aspects of temperature and precipitation. Soils in very cold areas with permafrost conditions tend to be shallow and weakly developed due to the short growing season. Organic rich surface horizons are common in low-lying areas due to limited decomposition. In warm, tropical soils, soils tend to be thicker, with extensive leaching and mineral alteration. In such climates, organic matter decomposition and chemical weathering occur at an accelerated rate.

Organisms

Animals, plants, and microorganisms all have important roles in soil development processes, in providing a supply of organic matter, and/or in nutrient cycling.

Worms, nematodes, termites, ants, gophers, moles, etc. all cause considerable mixing of soil and help to blend soil, aerate and lighten the soil by creating pores (which help store water and air).

Plant life provides organic matter to soil and helps to recycle nutrients with uptake by roots in the subsurface. The type of plant life that occurs in a given area, such as types of trees or grasses, depends on the climate, along with parent material and soil type. With the annual dropping of leaves and needles, trees tend to add organic matter to soil surfaces, helping to create a thin, organic-rich A or O horizon over time. Grasses, on the other hand, have a considerable root and surface masses that add to the soil each fall for annuals and short-lived perennials. For this reason, grassland soils have much thicker A horizons with higher organic matter contents, and are more agriculturally productive than forest soils.

Relief: Topography and Drainage

The local landscape can have a surprisingly strong effect on the soils that form on site. The local topography (**relief**) can have important microclimatic effects as well as affecting rates of soil erosion. In comparison to flat regions, areas with steep slopes overall have more soil erosion, more runoff of rainwater, and less water infiltration, all of which lead to more limited soil development in very hilly or mountainous areas. In the northern hemisphere, south-facing slopes are exposed to more direct sunlight angles and are thus warmer and drier than north-facing slopes. The cooler, moister north-facing slopes have a more dynamic plant community due to less evapotranspiration and, consequently, experience less erosion because of plant rooting of soil and have thicker soil development.

Soil drainage affects organic matter accumulation and preservation, and local vegetation types. Well-drained soils, generally on hills or sideslopes, are more brownish or reddish due to conversion of ferrous iron (Fe^{2+}) to minerals with ferric (Fe^{3+}) iron. More poorly drained soils, in lowland, alluvial plains or upland depressions, tend more be more greyish, greenish-grey (gleyed), or dark colored, due to iron reduction (to Fe^{2+}) and accumulation and preservation of organic matter in areas tending towards anoxic. Areas with poor drainage also tend to be lowlands into which soil material may wash and accumulate from surrounding uplands, often resulting in overthickened A or O horizons. In contrast, steeply sloping areas in highlands may experience erosion and have thinner surface horizons.

Parent Material of a Soil

The **parent material** of a soil is the material from which the soil has developed, whether it be river sands, shoreline deposits, glacial deposits, or various types of bedrock. In youthful soils, the parent material has a clear connection to the soil type and has significant influence. Over time, as weathering processes deepen, mix, and alter the soil, the parent material becomes less recognizable as chemical, physical, and biological processes take their effect. The type of parent material may also affect the rapidity of soil development. Parent materials that are highly weatherable (such as volcanic ash) will transform more quickly into highly developed soils, whereas parent materials that are quartz-rich, for example, will take longer to develop. Parent materials also provide nutrients to plants and can affect soil internal drainage (e.g. clay is more impermeable than sand and impedes drainage).

Time

In general, soil profiles tend to become thicker (deeper), more developed, and more altered over time. However, the rate of change is greater for soils in youthful stages of development. The degree of soil alteration and deepening slows with time and at some point, after tens or hundreds of thousands of years, may approach an equilibrium condition where erosion and deepening (removals and additions) become balanced. Young soils (< 10,000 years old) are strongly influenced by parent material and typically develop horizons and character rapidly. Moderate age soils (roughly 10,000 to 500,000 years old) are slowing in profile development and deepening, and may begin to approach equilibrium conditions. Old soils (>500,000 years old) have generally reached their limit as far as soil horizonation and physical structure, but may continue to alter chemically or mineralogically.

Soil development is not always continual. Geologic events can rapidly bury soils (landslides, glacier advance, lake transgression), can cause removal or truncation of soils (rivers, shorelines) or can cause soil renewal with additions of slowly deposited sediment that add to the soil (wind or floodplain deposits). Biological mixing can sometimes cause soil regression, a reversal or bump in the road for the normal path of increasing development over time.

9.2 SOIL-PLANT INTERACTIONS

Soil plays a key role in plant growth. Beneficial aspects to plants include providing physical support, water, heat, nutrients, and oxygen (Figure 1). Mineral nutrients from the soil can dissolve in water and then become available to plants. Although many aspects of soil are beneficial to plants, excessively high levels of trace metals (either naturally occurring or anthropogenically added) or applied herbicides can be toxic to some plants.

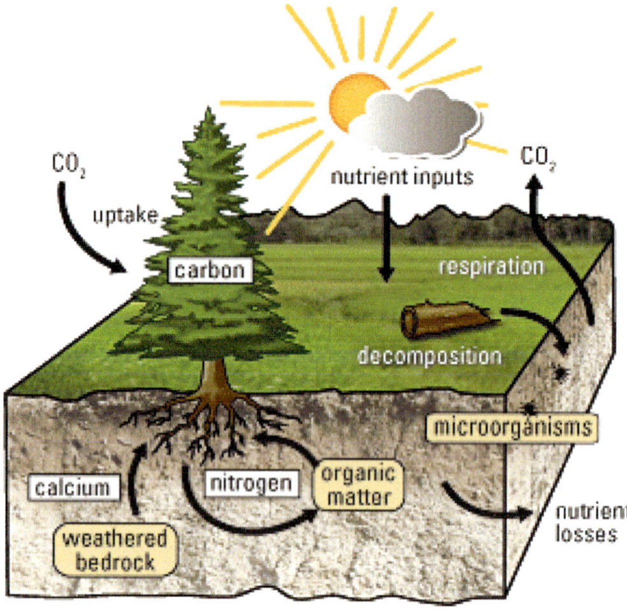

Figure 4. Soil-Plant Nutrient Cycle. This figure illustrates the uptake of nutrients by plants in the forest-soil ecosystem. Source: U.S. Geological Survey.

The ratio of solids/water/air in soil is also critically important to plants for proper oxygenation levels and water availability. Too much porosity with air space, such as in sandy or gravelly soils, can lead to less available water to plants, especially during dry seasons when the water table is low. Too much water, in poorly drained regions, can lead to anoxic conditions in the soil, which may be toxic to some plants.

Nutrient Uptake by Plants

Several elements obtained from soil are considered essential for plant growth. Macronutrients, including C, H, O, N, P, K, Ca, Mg, and S, are needed by plants in significant quantities. C, H, and O are mainly obtained from the atmosphere or from rainwater. These three elements are the main components of most organic compounds, such as proteins, lipids, carbohydrates, and nucleic acids. The other six elements (N, P, K, Ca, Mg, and S) are obtained by plant roots from the soil and are variously used for protein synthesis, chlorophyll synthesis, energy transfer, cell division, enzyme reactions, and homeostasis (the process regulating the conditions within an organism).

Micronutrients are essential elements that are needed only in small quantities, but can still be limiting to plant growth since these nutrients are not so abundant in nature. Micronutrients include iron (Fe), manganese (Mn), boron (B), molybdenum (Mo), chlorine (Cl), zinc (Zn), and copper (Cu). There are some other elements that tend to aid plant growth but are not absolutely essential.

Micronutrients and macronutrients are desirable in particular concentrations and can be detrimental to plant growth when concentrations in soil solution are either too low (limiting) or too high (toxicity). Mineral nutrients are useful to plants only if they are in an extractable form in soil solutions, such as a dissolved ion rather than in solid mineral. Many nutrients move through the soil and into the root system as a result of concentration gradients, moving by diffusion from high to low concentrations. However, some nutrients are selectively absorbed by the root membranes, enabling concentrations to become higher inside the plant than in the soil.

9.3 CONCEPTS OF CONVENTIONAL AGRICULTURE

The prevailing agricultural system, variously called "**conventional farming**," "modern agriculture," or "industrial farming," has delivered tremendous gains in productivity and efficiency. Food production worldwide has risen in the past 50 years; the World Bank estimates that between 70 percent and 90 percent of the recent increases in food production are the result of conventional agriculture rather than greater acreage under cultivation. U.S. consumers have come to expect abundant and inexpensive food.

Figure 5. Conventional agriculture is dependent on large investments in mechanized equipment powered mostly by fossil fuels. This has made agriculture efficient, but has had an impact on the environment. Cotton Harvest by Kimberly Vardeman is licensed under CC BY 4.0.

Conventional farming systems vary from farm to farm and from country to country. However, they share many characteristics such as rapid technological innovation, large capital investments in equipment and technology, large-scale farms, single crops (monocultures); uniform high-yield hybrid crops, dependency on agribusiness, mechanization of farm work, and extensive use of pesticides, fertilizers, and herbicides. In the case of livestock, most production comes from systems where animals are highly concentrated and confined.

Both positive and negative consequences have come with the bounty associated with industrial farming. Some concerns about conventional agriculture are presented below.

Ecological Concerns

Agriculture profoundly affects many ecological systems. Negative effects of current practices include the following:

Decline in soil productivity can be due to wind and water erosion of exposed topsoil, soil compaction, loss of soil organic matter, water holding capacity, and biological activity; and salinization (increased salinity) of soils in highly-irrigated farming areas. Converting land to desert (desertification) can be caused by overgrazing of livestock and is a growing problem, especially in parts of Africa.

Agricultural practices have been found to contribute to non-point source water pollutants that include salts, fertilizers (nitrates and phosphorus, especially), pesticides, and herbicides. Pesticides from every chemical class have been detected in groundwater and are commonly found in groundwater beneath agricultural areas. They are also widespread in the nation's surface waters. Eutrophication and "dead zones" due to nutrient runoff affect many rivers, lakes, and oceans. Reduced water quality impacts agricultural production, drinking water supplies, and fishery production. Water scarcity (discussed in the previous chapter) in many places is due to overuse of surface and ground water for irrigation with little concern for the natural cycle that maintains stable water availability.

Other environmental ills include over 400 insects and mite pests and more than 70 fungal pathogens that have become resistant to one or more pesticides. Pesticides have also placed stresses on pollinators and other beneficial insect species. This, along with habitat loss due to converting wildlands into agricultural fields, has affected entire ecosystems (such as the practice of converting tropical rainforests into grasslands for raising cattle).

Agriculture's link to global climate change is just beginning to be appreciated. Destruction of tropical forests and other native vegetation for agricultural production has a role in elevated levels of carbon dioxide and other greenhouse gases. Recent studies have found that soils may be large reservoirs of carbon.

Economic and Social Concerns

Economically, the U.S. agricultural sector includes a history of increasingly large federal expenditures. Also observed is a widening disparity among the income of farmers and the escalating concentration of **agribusiness**—industries involved with manufacture, processing, and distribution of farm products—into fewer and fewer hands. Market competition is limited and farmers have little control over prices of their goods, and they continue to receive a smaller and smaller portion of consumer dollars spent on agricultural products.

Economic pressures have led to a tremendous loss to the number of farms, particularly small farms, and farmers during the past few decades. More than 155,000 farms were lost from 1987 to 1997. Economically, it is very difficult for potential farmers to enter the business today because of the high cost of doing business. Productive farmland also has been swallowed up by urban and suburban sprawl—since 1970, over 30 million acres have been lost to development.

Health Hazards and Farming Practices

Many potential health hazards are tied to farming practices. The general public may be affected by the sub-therapeutic use of antibiotics in animal production and the contamination of food and water by pesticides and nitrates. These are areas of active research to determine the levels of risk. The health of farm workers is also of concern, as their risk of exposure is much higher.

Philosophical Considerations

Historically, farming played an important role in our development and identity as a nation. From strongly agrarian roots, we have evolved into a culture with few farmers. Less than two percent of Americans now produce food for all U.S. citizens. Can sustainable and equitable food production be established when most consumers have so little connection to the natural processes that produce their food? What intrinsically American values have changed and will change with the decline of rural life and farmland ownership?

World population continues to grow. According to recent United Nations population projections, the world population will grow to 9.7 billion in 2050 and 11.2 billion in 2100. The rate of population increase is especially high in many developing countries. In these countries, the population factor, combined with rapid industrialization, poverty, political instability, and large food imports and debt burden, make long-term food security especially urgent.

9.4 PESTS AND PESTICIDES

Pests are organisms that occur where they are not wanted or that cause damage to crops or humans or other animals. Thus, the term "pest" is a highly subjective term. A pesticide is a term for any substance intended for preventing, destroying, repelling, or mitigating any pest. Though often misunderstood to refer only to insecticides, the term pesticide also applies to herbicides, fungicides, and various other substances used to control pests. By their very nature, most pesticides create some risk of harm—pesticides can cause harm to humans, animals, and/or the environment because they are designed to kill or otherwise adversely affect living things. At the same time, pesticides are useful to society because they can kill potential disease-causing organisms and control insects, weeds, worms, and fungi.

Figure 6. The cotton boll weevil is considered a major pest because of the damage it does to cotton plants. This work by Jimmy Smith is licensed under CC BY-NC-ND 4.0.

Pest Control and Management

The management of pests is an essential part of agriculture, public health, and maintenance of power lines and roads. Chemical pest management has helped to reduce losses in agriculture and to limit human exposure to disease vectors, such as mosquitoes, saving many lives. Chemical pesticides can be effective, fast acting, adaptable to all crops and situations. When first applied, pesticides can result in impressive production gains of crops. However, despite these initial gains, excessive use of pesticides can be ecologically unsound, leading to the destruction of natural enemies, the increase of pesticide resistance, and outbreaks of secondary pests.

These consequences have often resulted in higher production costs as well as environmental and human health costs- side-effects which have been unevenly distributed. Despite the fact that the lion's share of chemical pesticides are applied in developed countries, 99 percent of all pesticide poisoning cases occur in developing countries where regulatory, health and education systems are weakest. Many farmers in developing countries overuse pesticides and do not take proper safety precautions because they do not understand the risks and fear smaller harvests. Making matters worse, developing countries seldom have strong regulatory systems for dangerous chemicals; pesticides banned or restricted in industrialized countries are used widely in developing countries. Farmers' perceptions of appropriate pesticide use vary by setting and culture. Prolonged exposure to pesticides has been associated with several chronic and acute health effects like non-Hodgkin's lymphoma, leukemia, as well as cardiopulmonary disorders, neurological and hematological symptoms, and skin diseases.

POPs

Persistent organic pollutants (POPs) are a group of organic chemicals, such as DDT, that have been widely used as pesticides or industrial chemicals and pose risks to human health and ecosystems. POPs have been produced and released into the environment by human activity. They have the following three characteristics:

Persistent: POPs are chemicals that last a long time in the environment. Some may resist breakdown for years and even decades while others could potentially break down into other toxic substances.

Bioaccumulative: POPs can accumulate in animals and humans, usually in fatty tissues and largely from the food they consume. As these compounds move up the food chain, they concentrate to levels that could be thousands of times higher than acceptable limits.

Toxic: POPs can cause a wide range of health effects in humans, wildlife and fish. They have been linked to effects on the nervous system, reproductive and developmental problems, suppression of the immune system, cancer, and endocrine disruption. The deliberate production and use of most POPs has been banned around the world, with some exemptions made for human health considerations (e.g., DDT for malaria control) and/or in very specific cases where alternative chemicals have not been identified. However, the unintended production and/or the current use of some POPs continue to be an issue of global concern. Even though most POPs have not been manufactured or used for decades, they continue to be present in the environment and thus potentially harmful. The same properties that originally made them so effective, particularly their stability, make them difficult to eradicate from the environment.

POPs and Human Health

The relationship between exposure to environmental contaminants such as POPs and human health is complex. There is mounting evidence that these persistent, bioaccumulative and toxic chemicals (PBTs) cause long-term harm to human health and the environment. Drawing a direct link, however, between exposure to these chemicals and health effects is complicated, particularly since humans are exposed on a daily basis to many different environmental contaminants through the air they breathe, the water they drink, and the food they eat. Numerous studies link POPs with a number of adverse effects in humans. These include effects on the nervous system, problems related to reproduction and development, cancer, and genetic impacts. Moreover, there is mounting public concern over the environmental contaminants that mimic hormones in the human body (**endocrine disruptors**).

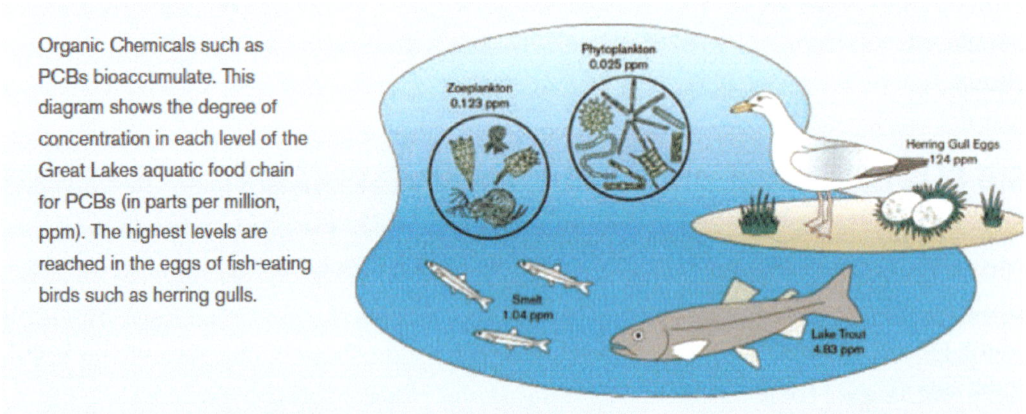

Figure 7. Bioaccumulation and biomagnification. U.S. EPA. Great Lakes: The Great Lakes Atlas: Chapter Four the Great Lakes Today – Concerns. January 2009

"Grasshopper Effect"

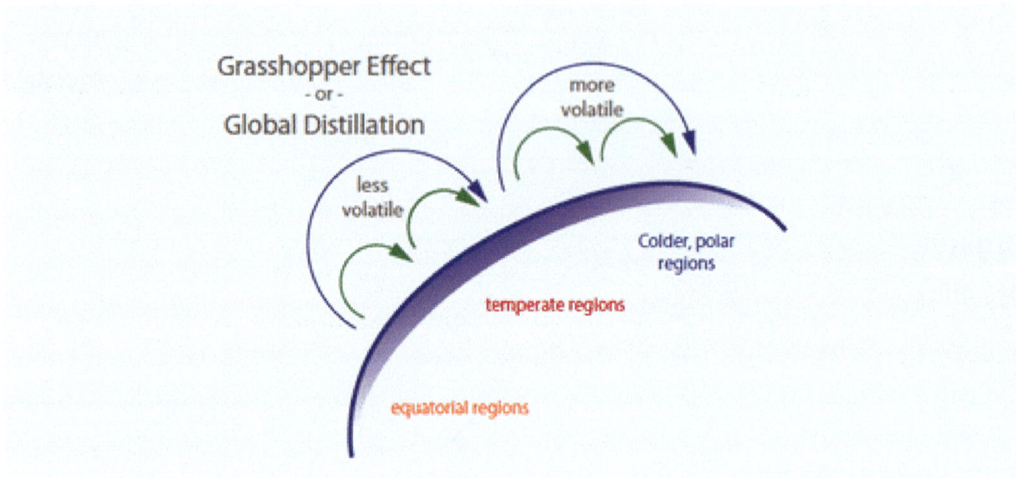

Figure 8. How POPs move throughout the environment. Source: Environment Canada. The Science and the Environment Bulletin. May/June 1998

While generally banned or restricted, POPs make their way into and throughout the environment on a daily basis through a cycle of long-range air transport and deposition called the "grasshopper effect."The "grasshopper" processes, illustrated in Figure 3, begin with the release of POPs into the environment. When POPs enter the atmosphere, they can be carried with wind currents, sometimes for long distances.

Through atmospheric processes, they are deposited onto land or into water ecosystems where they accumulate and potentially cause damage. From these ecosystems, they evaporate, again entering the atmosphere, typically traveling from warmer temperatures toward cooler regions. They condense out of the atmosphere whenever the temperature drops, eventually reaching highest concentrations in circumpolar countries. Through these processes, POPs can move thousands of kilometers from their original source of release in a cycle that may last decades.

9.5 SUSTAINABLE AGRICULTURE: AN INTEGRATED SYSTEM OF PLANT AND ANIMAL PRODUCTION PRACTICES

"**Sustainable agriculture**" was addressed by Congress in the 1990 "Farm Bill". Under that law, "the term sustainable agriculture means an integrated system of plant and animal production practices having a site-specific application that will, over the long term:

- satisfy human food and fiber needs;
- enhance environmental quality and the natural resource base upon which the agricultural economy depends;
- make the most efficient use of nonrenewable resources and on-farm resources and integrate, where appropriate, natural biological cycles and controls;
- sustain the economic viability of farm operations; and
- enhance the quality of life for farmers and society as a whole."

Organic Farming is Eco-friendly

Organic agriculture is an ecological production management system that promotes and enhances biodiversity, biological cycles and soil biological activity. Organic food is produced by farmers who emphasize the use of renewable resources and the conservation of soil and water to enhance environmental quality for future generations. Organic meat, poultry, eggs, and dairy products come from animals that are given no antibiotics or growth hormones. Organic food is produced without using most conventional pesticides, fertilizers made with synthetic ingredients or sewage sludge, or GMOs.

Organic production, with the corresponding practices to maintain soil fertility and soil health is therefore a more benign alternative to conventional, high-value horticulture. The organic food movement has been endorsed by the UN's Food and Agricultural Organization, which maintains in a 2007 report that organic farming fights hunger, tackles climate change, and is good for farmers, consumers, and the environment. The strongest benefits of organic agriculture are its use of resources that are independent of fossil fuels, are locally available, incur minimal environmental stresses, and are cost effective.

Integrated Pest Management: Combination of Common-Sense Practices

Integrated Pest Management (IPM) refers to a mix of farmer-driven, ecologically-based pest control practices that seeks to reduce reliance on synthetic chemical pesticides. It involves (a) managing pests (keeping them below economically damaging levels) rather than seeking to eradicate them; (b) relying, to the extent possible, on non-chemical measures to keep pest populations low; and (c) selecting and applying pesticides, when they have to be used, in a way that minimizes adverse effects on beneficial organisms, humans, and the environment. It is commonly understood that applying an IPM approach does not necessarily mean eliminating pesticide use, although this is often the case because pesticides are often over-used for a variety of reasons.

The IPM approach regards pesticides as mainly short-term corrective measures when more ecologically based control measures are not working adequately (sometimes referred to as using pesticides as the "last resort"). In those cases when pesticides are used, they should be selected and applied in such a manner as to minimize the amount of disruption that they cause to the environment, such as using products that are non-persistent and applying them in the most targeted way possible).

Biological Control

Figure 9. Young larvae of the leaf-beetle feed in and on the developing buds of plants, often destroying them. This may stunt plant growth and delay or prevent flowering. Adults (shown) and older larvae feed on leaves and cause severe defoliation. Leaf-beetles can be used to as a biocontrol for invasive plants such as purple loosestrife.

Biological control (biocontrol) is the use of one biological species to reduce populations of a different species. There has been a substantial increase in commercialization of biocontrol products, such as beneficial insects, cultivated predators and natural or non-toxic pest control products. Biocontrol is being mainstreamed to major agricultural commodities, such as cotton, corn and most commonly vegetable crops. Biocontrol is also slowly emerging in vector control in public health and in areas that for a long time mainly focused on chemical vector control in mosquito/malaria—and black fly/onchocerciasis—control programs. Successful and commercialized examples of biocontrol include ladybugs to depress aphid populations, parasitic wasps to reduce moth populations, use of the bacterium *Bacillus thuringenensis* to kill mosquito and moth larvae, and introduction of fungi, such as Trichoderma, to suppress fungal-caused plant diseases, leaf beetle (*Galerucella calmariensis*) to suppress purple loosestrife, a noxious weed (Figure 1). In all of these cases, the idea is not to completely destroy the pathogen or pest, but rather to reduce the damage below economically significant values.

Intercropping

Figure 10. Intercropping alyssum (the flowering plants shown center left) with organic romaine lettuce for aphid control.

Intercropping means growing two or more crops in close proximity to each other during part or all of their life cycles to promote soil improvement, biodiversity, and pest management. Incorporating intercropping principles into an agricultural operation increases diversity and interaction between plants, arthropods, mammals, birds and microorganisms resulting in a more stable crop-ecosystem and a more efficient use of space, water, sunlight, and nutrients (Figure 2). This collaborative type of crop management mimics nature and is subject to fewer pest outbreaks, improved nutrient cycling and crop nutrient uptake, and increased water infiltration and moisture retention. Soil quality, water quality and wildlife habitat all benefit.

Organic Farming Practices and Benefits

In modern agricultural practices, heavy machinery is used to prepare the seedbed for planting, to control weeds, and to harvest the crop. The use of heavy equipment has many advantages in saving time and labor, but can cause compaction of soil and disruption of the natural soil organisms. The problem with soil compaction is that increased soil density limits root penetration depth and may inhibit proper plant growth.

Alternative practices generally encourage minimal tillage or no tillage methods. With proper planning, this can simultaneously limit compaction, protect soil organisms, reduce costs (if performed correctly), promote water infiltration, and help to prevent topsoil erosion (Figure 3).

Figure 11. Farmers should consider no-till farming as the most important tool to prevent loss of soil moisture.

Tillage of fields does help to break up clods that were previously compacted, so best practices may vary at sites with different soil textures and composition. Another aspect of soil tillage is that it may lead to more rapid decomposition of organic matter due to greater soil aeration. Over large areas of farmland, this has the unintended consequence of releasing more carbon and nitrous oxides (greenhouse gases) into the atmosphere, thereby contributing to global warming effects. In no-till farming, carbon can actually become sequestered into the soil. Thus, no-till farming may be advantageous to sustainability issues on the local scale and the global scale. No-till systems of conservation farming have proved a major success in Latin America and are being used in South Asia and Africa.

Crop Rotation: Improving Soil Nutrients and Breaking Crop Pest Cycles

Crop rotations are planned sequences of crops over time on the same field. Rotating crops provides productivity benefits by improving soil nutrient levels and breaking crop pest cycles. Farmers may also choose to rotate crops in order to reduce their production risk through diversification or to manage scarce resources, such as labor, during planting and harvesting timing. This strategy reduces the pesticide costs by naturally breaking the cycle of weeds, insects and diseases. Also, grass and legumes in a rotation protect water quality by preventing excess nutrients or chemicals from entering water supplies.

An Alternative To Spraying: Bollworm Control In Shandong

- Farmers in Shandong (China) have been using innovative methods to control bollworm infestation in cotton when this insect became resistant to most pesticides. Among the control measures implemented were:
- The use of pest resistant cultivars and interplanting of cotton with wheat or maize.
- Use of lamps and poplar twigs to trap and kill adults to lessen the number of adults.
- If pesticides were used, they were applied on parts of cotton plant's stem rather than by spraying the whole field (to protect natural enemies of the bollworm).
- These and some additional biological control tools have proved to be effective in controlling insect populations and insect resistance, protecting surroundings and lowering costs.

The Prospects of Sustainable Agriculture

Many in the agricultural community have adopted the sense of urgency and direction pointed to by the sustainable agriculture concept. Sustainability has become an integral component of many government, commercial, and non-profit agriculture research efforts, and it is beginning to be woven into agricultural policy. Increasing numbers of farmers and ranchers have embarked on their own paths to sustainability, incorporating integrated and innovative approaches into their own enterprises.

AT AGLANCE

In agriculture and horticulture, soil generally refers to the medium for plant growth, typically material within the upper meter or two. Soil plays a key role in plant growth. Beneficial aspects to plants include providing physical support, heat, water, nutrients, and oxygen. Heat, light, and oxygen are also obtained by the atmosphere, but the roots of many plants also require oxygen. The prevailing agricultural system has delivered tremendous gains in productivity and efficiency. Food production worldwide has risen in the past 50 years. On the other hand, agriculture profoundly affects many ecological systems. Negative effects of current practices include ecological concerns, economic and social concerns and human health concerns. Pesticides from every chemical class have been detected in groundwater and are commonly found in groundwater beneath agricultural areas. Despite impressive production gains, excessive use of pesticides has proven to be ecologically unsound, leading to the destruction of natural enemies, the increase of pest resistance pest resurgence and outbreaks of secondary pests. These consequences have often resulted in higher production costs and lost markets due to undesirable pesticide residue levels, as well as environmental and human health costs. Alternative and sustainable practices in farming and land use include organic agriculture, integrated pest management and biological control.

CHAPTER-10

AIR POLLUTION, GLOBAL CLIMATE CHANGE & OZONE LAYER DEPLETION

10.1 ATMOSPHERIC POLLUTION: AN OVERVIEW

Air pollution occurs in many forms but can generally be thought of as gaseous and particulate contaminants that are present in the earth's atmosphere. Chemicals discharged into the air that have a direct impact on the environment are called **primary pollutants**. These primary pollutants sometimes react with other chemicals in the air to produce **secondary pollutants**.

Air pollution is typically separated into two categories: outdoor air pollution and indoor air pollution. **Outdoor air pollution** involves exposures that take place outside of the built environment. Examples include fine particles produced by the burning of coal, noxious gases such as sulfur dioxide, nitrogen oxides and carbon monoxide; ground-level ozone and tobacco smoke. **Indoor air pollution** involves exposures to particulates, carbon oxides, and other pollutants carried by indoor air or dust. Examples include household products and chemicals, out-gassing of building materials, allergens (cockroach and mouse dropping, mold, pollen), and tobacco smoke.

Major Sources of Air Pollution

A **stationary source** of air pollution refers to an emission source that does not move, also known as a point source. Stationary sources include factories, power plants, and dry cleaners. The term **area source** is used to describe many small sources of air pollution located together whose individual emissions may be below thresholds of concern, but whose collective emissions can be significant. Residential wood burners are a good example of a small source, but when combined with many other small sources, they can contribute to local and regional air pollution levels. Area sources can also be thought of as non-point sources, such as construction of housing developments, dry lake beds, and landfills.

A **mobile source** of air pollution refers to a source that is capable of moving under its own power. In general, mobile sources imply "on-road" transportation, which includes vehicles such as cars, sport utility vehicles, and buses. In addition, there is also a "non-road" or "off-road" category that includes gas-powered lawn tools and mowers, farm and construction equipment, recreational vehicles, boats, planes, and trains.

Agricultural sources arise from operations that raise animals and grow crops, which can generate emissions of gases and particulate matter. For example, animals confined to a barn or restricted area produce large amounts of manure. Manure emits various gases, particularly ammonia into the air. This ammonia can be emitted from the animal houses, manure storage areas, or from the land after the manure is applied. In crop production, the misapplication of fertilizers, herbicides, and pesticides can potentially result in aerial drift of these materials and harm may be caused.

Unlike the above mentioned sources of air pollution, air pollution caused by **natural sources** is not caused by people or their activities. An erupting volcano emits particulate matter and gases, forest and prairie fires can emit large quantities of "pollutants", dust storms can create large amounts of particulate matter, and plants and trees naturally emit volatile organic compounds which can form aerosols that can cause a natural blue haze. Wild animals in their natural habitat are also considered natural sources of "pollution".

Common Air Pollutants

The most commonly found air pollutants are *particulate matter, ground-level ozone, carbon monoxide, sulfur oxides, nitrogen oxides,* and *lead.* These pollutants can harm health and the environment, and cause property damage. Of the six pollutants, particle pollution and ground-level ozone are the most widespread health threats. The U.S. Environmental Protection Agency (EPA) regulates them by developing criteria based on considerations of human and environmental health.

Ground-level ozone is not emitted directly into the air, but is created by chemical reactions between oxides of nitrogen (NOx) and volatile organic compounds (VOC) in the presence of sunlight. Emissions from industrial facilities and electric utilities, motor vehicle exhaust, gasoline vapors, and chemical solvents are some of the major sources of NOx and VOC. Breathing ozone can trigger a variety of health problems, particularly for children, the elderly, and people of all ages who have lung diseases such as asthma. Ground level ozone can also have harmful effects on sensitive vegetation and ecosystems. (Ground-level ozone should not be confused with the ozone layer, which is high in the atmosphere and protects Earth from ultraviolet light; ground-level ozone provides no such protection).

Particulate matter, also known as particle pollution, is a complex mixture of extremely small particles and liquid droplets. Particle pollution is made up of a number of components, including acids (such as nitrates and sulfates), organic chemicals, metals, and soil or dust particles. The size of particles is directly linked to their potential for causing health problems. EPA is concerned about particles that are 10 micrometers in diameter or smaller because those are the particles that generally pass through the throat and nose and enter the lungs. Once inhaled, these particles can affect the heart and lungs and cause serious health effects.

Carbon monoxide (CO) is a colorless, odorless gas emitted from combustion processes. Nationally and, particularly in urban areas, the majority of CO emissions to ambient air come from mobile sources. CO can cause harmful health effects by reducing oxygen delivery to the body's organs (like the heart and brain) and tissues. At extremely high levels, CO can cause death.

Nitrogen dioxide (NO2) is one of a group of highly reactive gasses known as "oxides of nitrogen," or nitrogen oxides (NOx). Other nitrogen oxides include nitrous acid and nitric acid. EPA's National Ambient Air Quality Standard uses NO2 as the indicator for the larger group of nitrogen oxides. NO2 forms quickly from emissions from cars, trucks and buses, power plants, and off-road equipment. In addition to contributing to the formation of ground-level ozone, and fine particle pollution, NO2 is linked with a number of adverse effects on the respiratory system.

Sulfur dioxide (SO2) is one of a group of highly reactive gasses known as "oxides of sulfur." The largest sources of SO2 emissions are from fossil fuel combustion at power plants (73%) and other industrial facilities (20%). Smaller sources of SO2 emissions include industrial processes such as extracting metal from ore, and the burning of high sulfur containing fuels by locomotives, large ships, and non-road equipment. SO2 is linked with a number of adverse effects on the respiratory system.

Lead is a metal found naturally in the environment as well as in manufactured products. The major sources of lead emissions have historically been from fuels in on-road motor vehicles (such as cars and trucks) and industrial sources. As a result of regulatory efforts in the U.S. to remove lead from on-road motor vehicle gasoline, emissions of lead from the transportation sector dramatically declined by 95 percent between 1980 and 1999, and levels of lead in the air decreased by 94 percent between 1980 and 1999. Today, the highest levels of lead in air are usually found near lead smelters. The major sources of lead emissions to the air today are ore and metals processing and piston-engine aircraft operating on leaded aviation gasoline.

Indoor Air Pollution

Most people spend approximately 90 percent of their time indoors. However, the indoor air we breathe in homes and other buildings can be more polluted than outdoor air and can increase the risk of illness. There are many sources of indoor air pollution in homes. They include biological contaminants such as bacteria, molds and pollen, burning of fuels and environmental tobacco smoke, building materials and furnishings, household products, central heating and cooling systems, and outdoor sources. Outdoor air pollution can enter buildings and become a source of indoor air pollution.

Sick building syndrome is a term used to describe situations in which building occupants have health symptoms that are associated only with spending time in that building. Causes of sick building syndrome are believed to include inadequate ventilation, indoor air pollution, and biological contaminants. Usually indoor air quality problems only cause discomfort. Most people feel better as soon as they remove the source of the pollution. Making sure that your building is well-ventilated and getting rid of pollutants can improve the quality of your indoor air.

Secondhand Smoke

- Secondhand smoke is the combination of smoke that comes from a cigarette and smoke breathed out by a smoker. When a non-smoker is around someone smoking, they breathe in secondhand smoke.
- Secondhand smoke is dangerous to anyone who breathes it in. There is no safe amount of secondhand smoke. It contains over 7,000 harmful chemicals, at least 250 of which are known to damage human health. It can also stay in the air for several hours after somebody smokes. Even breathing secondhand smoke for a short amount of time can hurt your body.
- Over time, secondhand smoke can cause serious health issues in non-smokers. The only way to fully protect non-smokers from the dangers of secondhand smoke is to not allow smoking indoors. Separating smokers from nonsmokers (like "no smoking" sections in restaurants), cleaning the air, and airing out buildings does not completely get rid of secondhand smoke.

Source: Smokefree.gov

10.2 OZONE LAYER DEPLETION

The Ozone Layer

Ozone (O3) is a gaseous molecule that occurs in different parts of the atmosphere (Figure 1). It is very reactive chemically and is dangerous to plant and animal life when present in the lower portions of the atmosphere. This type of ozone, called ground-level ozone, is a significant hazard to human health and is associated with pollution from vehicle exhaust and other anthropogenic emissions (see section 10.1).

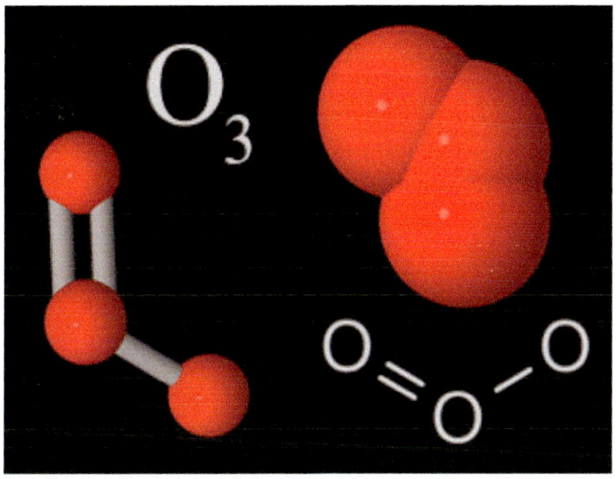

Figure 1. The ozone molecule is comprised of three oxygen atoms covalently bonded. "Ozone" by UCAR (Randy Russell) is licensed under CC BY-SA.

Ozone that occurs in the upper atmosphere is naturally occurring and beneficial to life due to its role in blocking harmful radiation from the sun. This type of ozone is called stratospheric ozone. Ozone in the stratosphere (Figure 2) forms when the two oxygen atoms in an O2 molecule are broken apart by the energy of sunlight. Each lone oxygen atom can then combine with a different O2 molecule to form O3, ozone. The ozone layer is the portion of the stratosphere where ozone molecules are present, mixed in among the other gases that comprise the atmosphere (Figure 2).

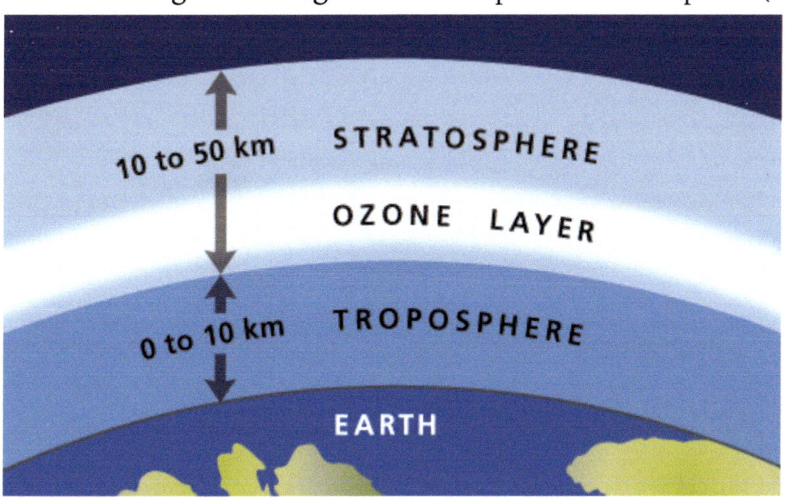

Figure 2. The ozone layer occurs in the region of the atmosphere called the stratosphere. This Photo by NC Climate Office is licensed under CC BY-SA.

Radiation from the sun is also called electromagnetic radiation, or simply referred to as light. The sun emits different types of light including but not limited to x-rays, visible light, microwaves, and ultraviolet light. The various types of light are distinguished by their different wavelengths. As wavelength decreases, the amount of energy in that light increases. Ultraviolet light, for example, has shorter wavelengths than visible light and is thus more energetic. Ozone molecules absorb ultraviolet (UV) light, which is advantageous for life on Earth because UV light can break down important biomolecules such as DNA, which can lead to cell death and mutations.

Ozone Layer Depletion

Unfortunately, the ozone layer that protects life on Earth from harmful UV light has been depleted due to human activities. The ozone depletion process begins when CFCs (chlorofluorocarbons) and other ozone-depleting substances (ODS) are emitted into the atmosphere. CFCs were used by industry as refrigerants, degreasing solvents, and propellants. In the lower atmosphere, CFC molecules are extremely stable chemically and do not dissolve in rain, and thus can linger for long periods. After several years, ODS molecules eventually reach the ozone layer in the stratosphere, which starts at about 10 kilometers above the Earth's surface.

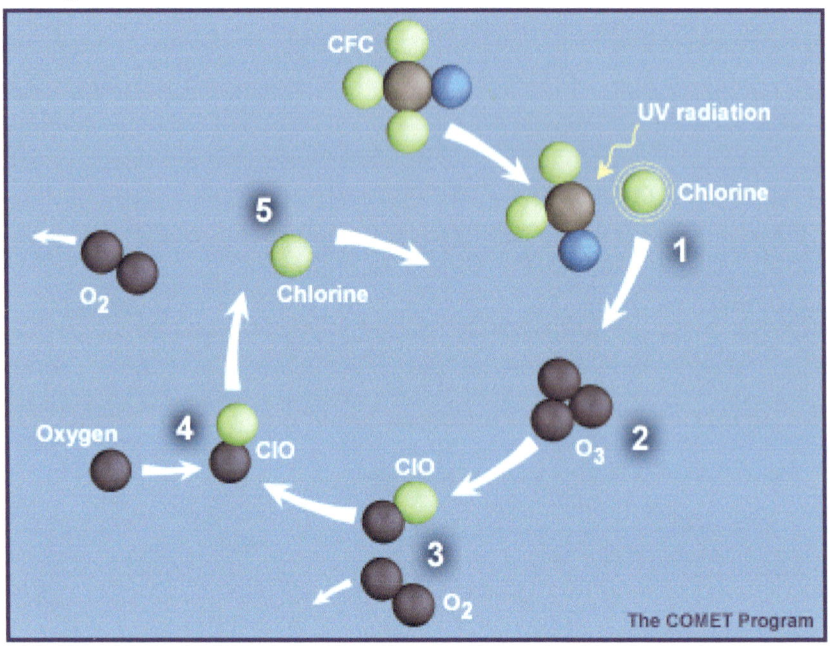

Figure 3. CFCs destroy ozone (O_3) when a chlorine atom breaks lose from the CFC and interacts with the oxygen atoms in the ozone molecule. This Photo by UCAR is licensed under CC BY-SA-NC.

Once in the stratosphere, CFCs and other ODS destroy ozone molecules. In the case of CFCs, UV light in the stratosphere knocks loose a chlorine atom from the molecule, which can then destroy numerous ozone molecules, as shown in Figure 3.

In effect, ODS are removing ozone faster than it is created by natural processes (as described above), and this leads to a thinning of the ozone layer. This thinning represents a reduction in concentration of ozone molecules in a particular portion of the stratosphere. Areas where the ozone layer has thinned are commonly referred to as holes, although this is not entirely accurate because ozone is still present, it just exists at concentrations much lower than normal.

Major Policies to Reduce Ozone Destruction

Tackling the issue of ozone layer destruction is an example of global cooperation that produced meaningful action on a large-scale environmental problem. In 1973, scientists first calculated that CFCs could reach the stratosphere and destroy ozone. Based only on their calculations, the United States and most Scandinavian countries banned CFCs in spray cans in 1978. But more confirmation that CFCs break down ozone was needed before additional action was taken. In 1985, members of the British Antarctic Survey reported that a 50% reduction in the ozone layer had been found over Antarctica in the previous three springs, a very important finding. Two years after that seminal British Antarctic Survey report, an agreement titled the "Montreal Protocol on Substances that Deplete the Ozone Layer" was ratified by nations all over the world. The **Montreal Protocol,** as it is commonly called, controls the production and emission of 96 chemicals that damage the ozone layer. As a result, CFCs have been mostly phased out since 1995, although they were used in developing nations until 2010. Some of the less hazardous substances will not be phased out until 2030. The Montreal Protocol also requires that wealthier nations donate money to develop technologies that will replace these chemicals.

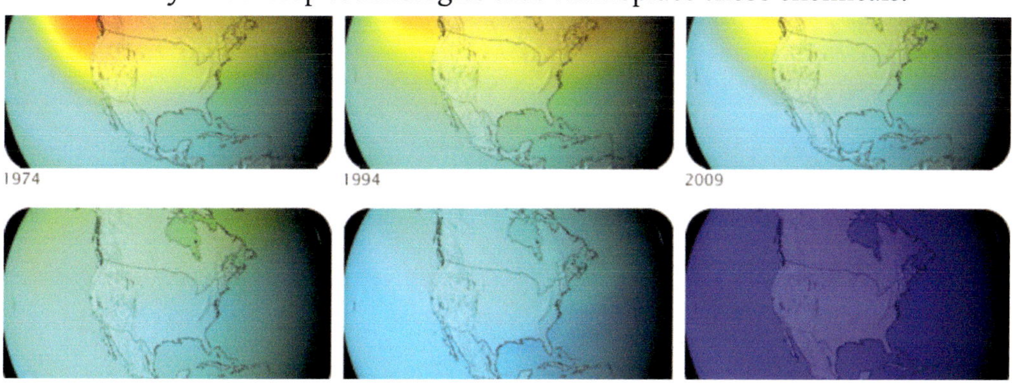

Figure 4. Ozone levels over North America decreased between 1974 and 2009. Models of the future predict what ozone levels would have been if CFCs were not being phased out. Warmer colors indicate more ozone.

The Montreal Protocol was a success and scientists have found that the ozone layer is recovering and the size of the ozone "holes" are shrinking, thanks to a drastic reduction in emission of ODS like CFCs. The recovery process is slow, however, because CFCs take many years to reach the stratosphere and can survive there a long time before they break down and are rendered harmless. Thus, it will take many more decades for the ozone layer to fully recover.

Constant vigilance and monitoring are needed, however, as illegal production and emission of CFCs and other ODS threaten recovery efforts. In 2018, scientists from the US National Oceanic and Atmospheric Administration reported that emissions of a particular type of CFC had increased 25% since 2012. Follow-up studies have since approximated the emissions as originating in particular regions of eastern Asia.

Ozone Layer Depletion and its Effect on Human Health and Environment

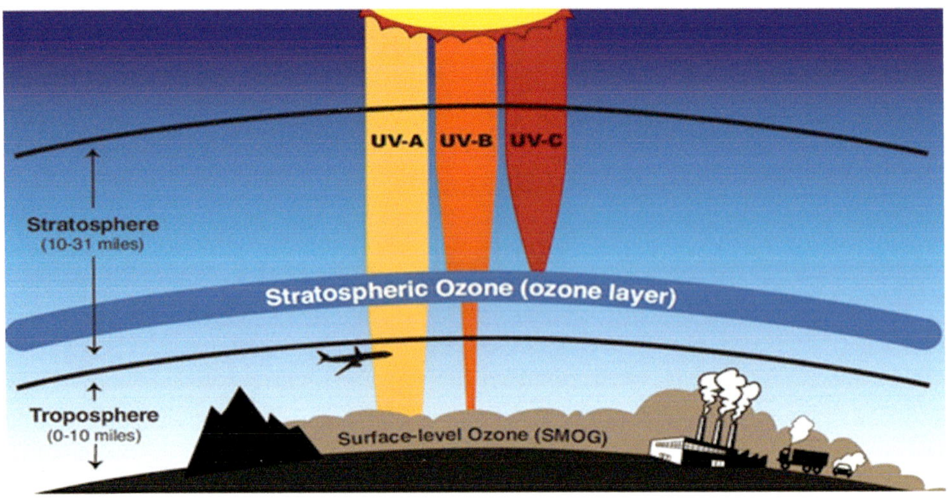

Figure 5. The ozone layer absorbs UV-B and UV-C light, protecting life on Earth from its harmful effects. This Photo by NASA is licensed under CCO Public Domain.

There are three types of UV light, each distinguished by their wavelengths: UV-A, UV-B, and UV-C. Stratospheric ozone molecules absorb all of the sun's UV-C light and most of its UV-B light (Figure 5).

Reductions in stratospheric ozone levels led to higher levels of UV-B reaching the Earth's surface, which is a serious hazard to human health. Studies have shown that in the Antarctic the amount of UV-B measured at the surface can double due to thinning of the ozone layer. UV-B is harmful to cells because it can interact with biomolecules like DNA and damage it. This can lead to mutations and cell death. UV-B cannot penetrate multicellular organisms very far and thus tends to only affect cells near the surface, such as in the skin of animals. Microbes like bacteria, however, are composed of only one cell and can therefore be killed by UV-B.

Laboratory and epidemiological studies demonstrate that UV-B causes certain types of skin cancers in humans and plays a major role in development of malignant melanoma (a particularly dangerous form of skin cancer). In addition, UV-B causes cataracts, a clouding of the lens in the eye that can lead to poor vision or even blindness.

It is important to note that all sunlight contains some UV-B light, even with normal stratospheric ozone levels. Therefore, it is important to protect your skin and eyes from the sun. Ozone layer depletion increases the amount of UV-B and the risk of health effects.

10.3 ACID RAIN AND ITS EFFECTS

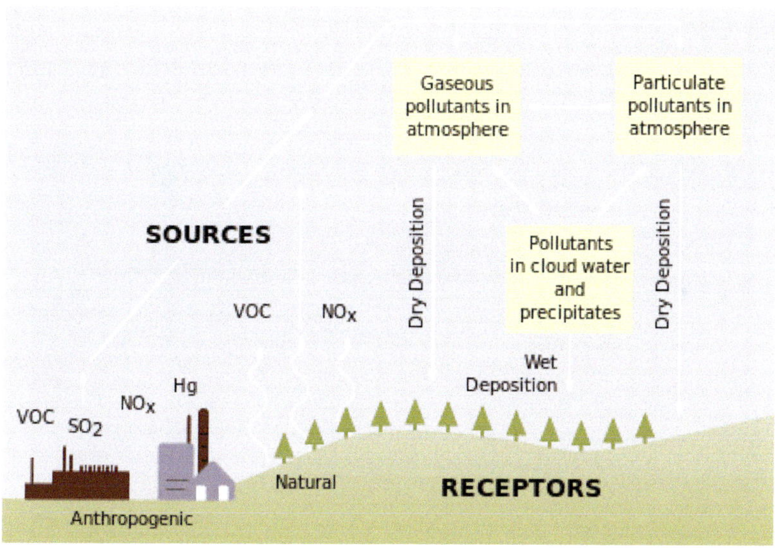

Figure 6. Processes involved in acid deposition.

Acid rain is a term referring to a mixture of wet and dry deposition (deposited material) from the atmosphere containing higher than normal amounts of nitric and sulfuric acids. The precursors, or chemical forerunners, of acid rain formation result from both natural sources, such as volcanoes and decaying vegetation, and man-made sources, primarily emissions of sulfur dioxide (SO_2) and nitrogen oxides (NO_x) resulting from fossil fuel combustion. Acid rain occurs when these gases react in the atmosphere with water, oxygen, and other chemicals to form various acidic compounds. The result is a mild solution of sulfuric acid and nitric acid. When sulfur dioxide and nitrogen oxides are released from power plants and other sources, prevailing winds blow these compounds across state and national borders, sometimes over hundreds of miles.

Acid rain is measured using a scale called "pH." The lower a substance's pH, the more acidic it is. Pure water has a pH of 7.0. However, normal rain is slightly acidic because carbon dioxide (CO_2) dissolves into it forming weak carbonic acid, giving the resulting mixture a pH of approximately 5.6 at typical atmospheric concentrations of CO_2. As of 2000, the most acidic rain falling in the U.S. has a pH of about 4.3.

Effects of Acid Rain

Acid rain causes acidification of lakes and streams and contributes to the damage of trees at high elevations (for example, red spruce trees above 2,000 feet) and many sensitive forest soils. In addition, acid rain accelerates the decay of building materials and paints, including irreplaceable buildings, statues, and sculptures that are part of our nation's cultural heritage. Prior to falling to the earth, sulfur dioxide (SO2) and nitrogen oxide (NOx) gases and their particulate matter derivatives—sulfates and nitrates—contribute to visibility degradation and harm public health.

Figure 7. A gargoyle that has been damaged by acid rain.

The ecological effects of acid rain are most clearly seen in the aquatic, or water, environments, such as streams, lakes, and marshes. Most lakes and streams have a pH between 6 and 8, although some lakes are naturally acidic even without the effects of acid rain. Acid rain primarily affects sensitive bodies of water, which are located in watersheds whose soils have a limited ability to neutralize acidic compounds (called "buffering capacity"). Lakes and streams become acidic (i.e., the pH value goes down) when the water itself and its surrounding soil cannot buffer the acid rain enough to neutralize it. In areas where buffering capacity is low, acid rain releases aluminum from soils into lakes and streams; aluminum is highly toxic to many species of aquatic organisms. Acid rain causes slower growth, injury, or death of forests. Of course, acid rain is not the only cause of such conditions. Other factors contribute to the overall stress of these areas, including air pollutants, insects, disease, drought, or very cold weather. In most cases, in fact, the impacts of acid rain on trees are due to the combined effects of acid rain and these other environmental stressors.

Acid rain and the dry deposition of acidic particles contribute to the corrosion of **metals** (such as bronze) and the deterioration of paint and stone (such as marble and limestone). These effects significantly reduce the societal value of buildings, bridges, cultural objects (such as statues, monuments, and tombstones), and cars (Figure 2).

Sulfates and nitrates that form in the atmosphere from sulfur dioxide (SO_2) and nitrogen oxides (NO_x) emissions contribute to **visibility impairment,** meaning we cannot see as far or as clearly through the air. The pollutants that cause acid rain—sulfur dioxide (SO_2) and nitrogen oxides (NO_x)—damage **human health.** These gases interact in the atmosphere to form fine sulfate and nitrate particles that can be transported long distances by winds and inhaled deep into people's lungs. Fine particles can also penetrate indoors. Many scientific studies have identified a relationship between elevated levels of fine particles and increased illness and premature death from heart and lung disorders, such as asthma and bronchitis.

10.4 GLOBAL CLIMATE CHANGE: CAUSES AND EFFECTS

Earth's Temperature is a Balancing Act

Earth's temperature depends on the balance between energy entering and leaving the planet. When incoming energy from the sun is absorbed, Earth warms. When the sun's energy is reflected back into space, Earth avoids warming. When energy is released from Earth into space, the planet cools. Many factors, both natural and human, can cause changes in Earth's energy balance, including:

- Changes in the greenhouse effect, which affects the amount of heat retained by Earth's atmosphere;
- Variations in the sun's energy reaching Earth;
- Changes in the reflectivity of Earth's atmosphere and surface.

Scientists have pieced together a picture of Earth's climate, dating back hundreds of thousands of years, by analyzing a number of indirect measures of climate such as ice cores, tree rings, glacier size, pollen counts, and ocean sediments. Scientists have also studied changes in Earth's orbit around the sun and the activity of the sun itself. The historical record shows that the climate varies naturally over a wide range of time scales. In general, climate changes prior to the Industrial Revolution in the 1700s can be explained by natural causes, such as changes in solar energy, volcanic eruptions, and natural changes in greenhouse gas (GHG) concentrations. Recent changes in climate, however, cannot be explained by natural causes alone. Research indicates that natural causes are very unlikely to explain most observed warming, especially warming since the mid-20th century. Rather, it is extremely likely that human activities, especially our combustion of fossil fuels, explains most of that warming. The scientific consensus is clear: through alterations of the carbon cycle, humans are changing the global climate by increasing the effects of something known as the greenhouse effect.

The Greenhouse Effect Causes the Atmosphere to Retain Heat

Gardeners that live in moderate or cool environments use greenhouses because they trap heat and create an environment that is warmer than outside temperatures. This is great for plants that like heat, or are sensitive to cold temperatures, such as tomato and pepper plants. Greenhouses contain glass or plastic that allow visible light from the sun to pass. This light, which is a form of energy, is absorbed by plants, soil, and surfaces and heats them. Some of that heat energy is then radiated outwards in the form of infrared radiation, a different form of energy. Unlike with visible light, the glass of the greenhouse blocks the infrared radiation, thereby trapping the heat energy, causing the temperature within the greenhouse to increase.

The same phenomenon happens inside a car on a sunny day. Have you ever noticed how much hotter a car can get compared to the outside temperature? Light energy from the sun passes through the windows and is absorbed by the surfaces in the car such as seats and the dashboard. Those warm surfaces then radiate infrared radiation, which cannot pass through the glass. This trapped infrared energy causes the air temperatures in the car to increase. This process is commonly known as the greenhouse effect.

The greenhouse effect also happens with the entire Earth. Of course, our planet is not surrounded by glass windows. Instead, the Earth is wrapped with an atmosphere that contains greenhouse gases (GHGs). Much like the glass in a greenhouse, GHGs allow incoming visible light energy from the sun to pass, but they block infrared radiation that is radiated from the Earth towards space (Figure 1). In this way, they help trap heat energy that subsequently raises air temperature. Being a greenhouse gas is a physical property of certain types of gases; because of their molecular structure they absorb wavelengths of infrared radiation, but are transparent to visible light. Some notable greenhouse gases are water vapor (H2O), carbon dioxide (CO2), and methane (CH4). GHGs act like a blanket, making Earth significantly warmer than it would otherwise be. Scientists estimate that average temperature on Earth would be -18º C without naturally-occurring GHGs.

Global Warming
Global warming refers to the recent and ongoing rise in global average temperature near Earth's surface. It is caused mostly by increasing concentrations of greenhouse gases in the atmosphere. Global warming is causing climate patterns to change. However, global warming itself represents only one aspect of climate change.
Climate Change
Climate change refers to any significant change in the measures of climate lasting for an extended period of time. In other words, climate change includes major changes in temperature, precipitation, or wind patterns, among other effects, that occur over several decades or longer.

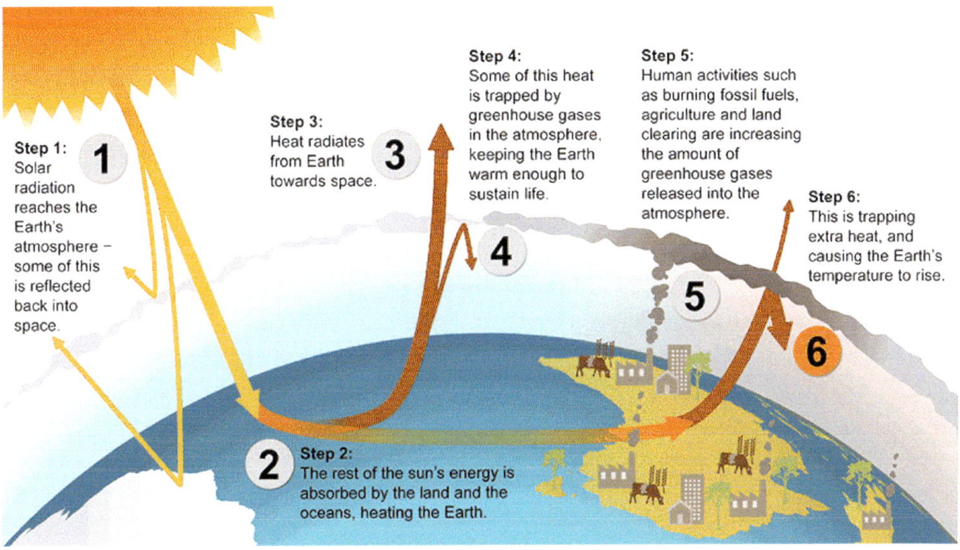

Figure 8. Enhanced Greenhouse Effect

Main Greenhouse Gasses

The most important GHGs directly emitted by humans include CO2 and methane. Carbon dioxide (CO2) is the primary greenhouse gas that is contributing to recent global climate change. CO2 is a natural component of the carbon cycle, involved in such activities as photosynthesis, respiration, volcanic eruptions, and ocean-atmosphere exchange. Human activities, primarily the burning of fossil fuels and changes in land use, release very large amounts of CO2 to the atmosphere, causing its concentration in the atmosphere to rise.

Atmospheric CO2 concentrations have increased by 45% since pre-industrial times, from approximately 280 parts per million (ppm) in the 18th century to 408 ppm in 2018. The current CO2 level is higher than it has been in at least 800,000 years, based on evidence from ice cores that preserve ancient atmospheric gases. Human activities currently release over 30 billion tons of CO2 into the atmosphere every year. While some some volcanic eruptions released large quantities of CO2 in the distant past, the U.S. Geological Survey (USGS) reports that human activities now emit more than 135 times as much CO2 as volcanoes each year. This human-caused build-up of CO2 in the atmosphere is like a tub filling with water, where more water flows from the faucet than the drain can take away.

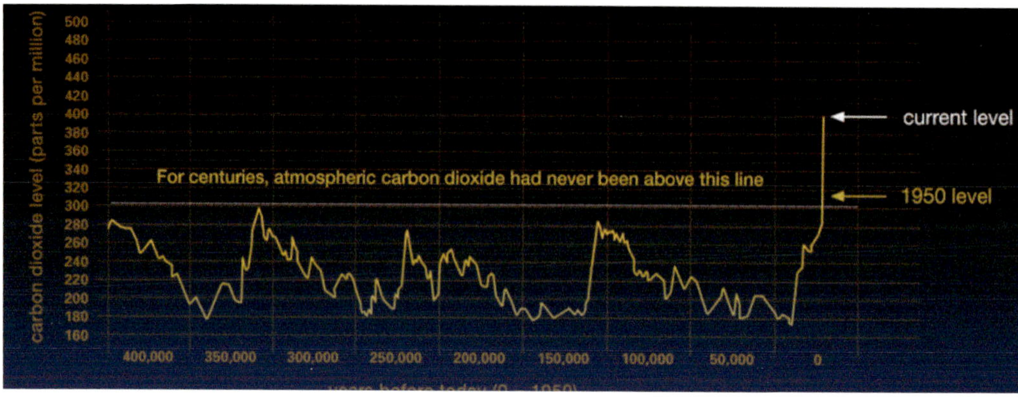

Figure 9. This graph, based on the comparison of atmospheric samples contained in ice cores and more recent direct measurements, provides evidence that atmospheric CO2 has increased since the Industrial Revolution. (Credit: Vostok ice core data/J.R. Petit et al.; NOAA Mauna Loa CO2 record.)

Methane (CH_4) is produced through both natural and human activities. For example, wetlands, agricultural activities, and fossil fuel extraction and transport all emit CH_4. Methane is more abundant in Earth's atmosphere now than at any time in at least the past 650,000 years. Due to human activities, CH_4 concentrations increased sharply during most of the 20th century and are now more than two and-a-half times pre-industrial levels. In recent decades, the rate of increase has slowed considerably.

Other Greenhouse Gasses

Water vapor is the most abundant greenhouse gas and also the most important in terms of its contribution to the natural greenhouse effect, despite having a short atmospheric lifetime. Some human activities can influence local water vapor levels. However, on a global scale, the concentration of water vapor is controlled by temperature, which influences overall rates of evaporation and precipitation. Therefore, the global concentration of water vapor is not substantially affected by direct human emissions.

Ground-level ozone (O3), which also has a short atmospheric lifetime, is a potent greenhouse gas. Chemical reactions create ozone from emissions of nitrogen oxides and volatile organic compounds from automobiles, power plants, and other industrial and commercial sources in the presence of sunlight (as discussed in section 10.1). In addition to trapping heat, ozone is a pollutant that can cause respiratory health problems and damage crops and ecosystems.

Changes in the Sun's Energy Affect how Much Energy Reaches Earth

Climate can be influenced by natural changes that affect how much solar energy reaches Earth. These changes include changes within the sun and changes in Earth's orbit. Changes occurring in the sun itself can affect the intensity of the sunlight that reaches Earth's surface. The intensity of the sunlight can cause either warming (during periods of stronger solar intensity) or cooling (during periods of weaker solar intensity). The sun follows a natural 11-year cycle of small ups and downs in intensity, but the effect on Earth's climate is small. Changes in the shape of Earth's orbit as well as the tilt and position of Earth's axis can also affect the amount of sunlight reaching Earth's surface.

Changes in the sun's intensity have influenced Earth's climate in the past. For example, the so-called "**Little Ice Age**" between the 17th and 19th centuries may have been partially caused by a low solar activity phase from 1645 to 1715, which coincided with cooler temperatures. The Little Ice Age refers to a slight cooling of North America, Europe, and probably other areas around the globe. Changes in Earth's orbit have had a big impact on climate over tens of thousands of years. These changes appear to be the primary cause of past cycles of ice ages, in which Earth has experienced long periods of cold temperatures (ice ages), as well as shorter interglacial periods (periods between ice ages) of relatively warmer temperatures.

Changes in solar energy continue to affect climate. However, solar activity has been relatively constant, aside from the 11-year cycle, since the mid-20th century and therefore does not explain the recent warming of Earth. Similarly, changes in the shape of Earth's orbit as well as the tilt and position of Earth's axis affect temperature on relatively long timescales (tens of thousands of years), and therefore cannot explain the recent warming.

Changes in Reflectivity Affect How Much Energy Enters Earth's System

When sunlight energy reaches Earth it can be reflected or absorbed. The amount that is reflected or absorbed depends on Earth's surface and atmosphere. Light-colored objects and surfaces, like snow and clouds, tend to reflect most sunlight, while darker objects and surfaces, like the ocean and forests, tend to absorb more sunlight. The term **albedo** refers to the amount of solar radiation reflected from an object or surface, often expressed as a percentage. Earth as a whole has an albedo of about %30, meaning that %70 of the sunlight that reaches the planet is absorbed. Sunlight that is absorbed warms Earth's land, water, and atmosphere. Albedo is also affected by aerosols. **Aerosols** are small particles or liquid droplets in the atmosphere that can absorb or reflect sunlight. Unlike greenhouse gases (GHGs), the climate effects of aerosols vary depending on what they are made of and where they are emitted. Those aerosols that reflect sunlight, such as particles from volcanic eruptions or sulfur emissions from burning coal, have a cooling effect. Those that absorb sunlight, such as black carbon (a part of soot), have a warming effect.

Natural changes in albedo, like the melting of sea ice or increases in cloud cover, have contributed to climate change in the past, often acting as feedbacks to other processes. Volcanoes have played a noticeable role in climate. Volcanic particles that reach the upper atmosphere can reflect enough sunlight back to space to cool the surface of the planet by a few tenths of a degree for several years. Volcanic particles from a single eruption do not produce long-term change because they remain in the atmosphere for a much shorter time than GHGs.

Human changes in land use and land cover have changed Earth's albedo. Processes such as deforestation, reforestation, desertification, and urbanization often contribute to changes in climate in the places they occur. These effects may be significant regionally, but are smaller when averaged over the entire globe.

Scientific Consensus: Global Climate Change is Real

The **Intergovernmental Panel on Climate Change** (IPCC) was created in 1988 by the United Nations Environment Programme and the World Meteorological Organization. It is charged with the task of evaluating and synthesizing the scientific evidence surrounding global climate change. The IPCC uses this information to evaluate current impacts and future risks, in addition to providing policymakers with assessments. These assessments are released about once every every six years. The most recent report, the 5th Assessment, was released in 2013. Hundreds of leading scientists from around the world are chosen to author these reports. Over the history of the IPCC, these scientists have reviewed thousands of peer-reviewed, publicly available studies. The scientific consensus is clear: global climate change is real and humans are very likely the cause for this change.

Additionally, the major scientific agencies of the United States, including the National Aeronautics and Space Administration (NASA) and the National Oceanic and Atmospheric Administration (NOAA), also agree that climate change is occurring and that humans are driving it. In 2010, the US National Research Council concluded that "Climate change is occurring, is very likely caused by human activities, and poses significant risks for a broad range of human and natural systems". Many independent scientific organizations have released similar statements, both in the United States and abroad. This doesn't necessarily mean that every scientist sees eye to eye on each component of the climate change problem, but broad agreement exists that climate change is happening and is primarily caused by excess greenhouse gases from human activities. Critics of climate change, driven by ideology instead of evidence, try to suggest to the public that there is no scientific consensus on global climate change. Such an assertion is patently false.

Global Climate Change: Future Changes

Greenhouse gas concentrations in the atmosphere will continue to increase unless the billions of tons of anthropogenic emissions each year decrease substantially. Increased concentrations are expected to:

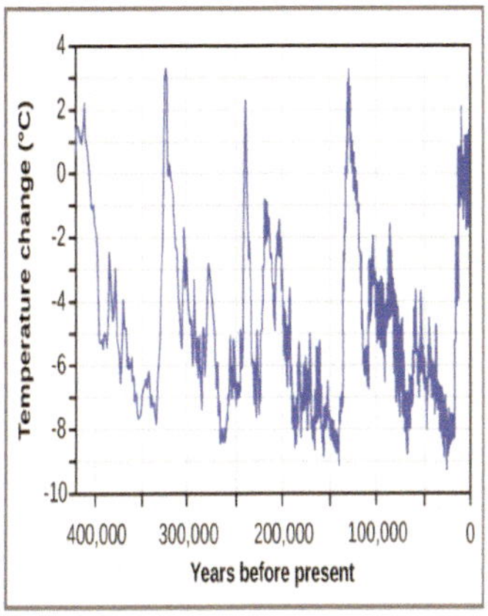

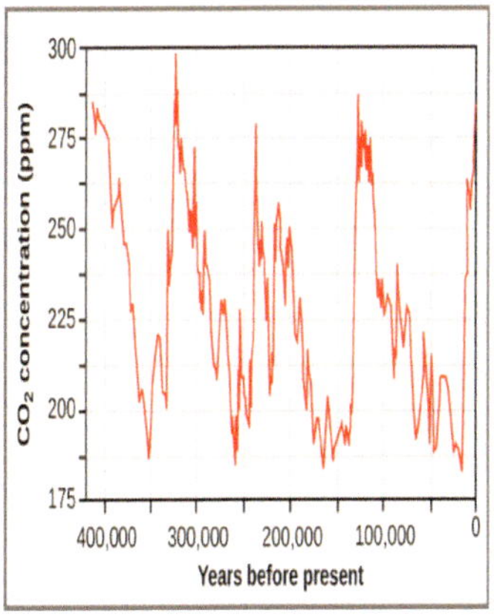

Figure 10. Ice at the Russian Vostok station in East Antarctica was laid down over the course 420,000 years and reached a depth of over 3,000 m. By measuring the amount of CO_2 trapped in the ice, scientists have determined past atmospheric CO_2 concentrations. Temperatures relative to modern day were determined from the amount of deuterium (an isotope of hydrogen) present.

- Increase Earth's average temperature,
- Influence the patterns and amounts of precipitation,

- Reduce ice and snow cover, as well as permafrost,
- Raise sea level,
- Increase the acidity of the oceans.

These changes will impact our food supply, water resources, infrastructure, ecosystems, and even our own health. The magnitude and rate of future climate change will primarily depend on the following factors:

- The rate at which levels of greenhouse gas concentrations in our atmosphere continue to increase,
- How strongly features of the climate (e.g., temperature, precipitation, and sea level) respond to the expected increase in greenhouse gas concentrations,
- Natural influences on climate (e.g., from volcanic activity and changes in the sun's intensity) and natural processes within the climate system (e.g., changes in ocean circulation patterns).

GHG Emissions Will Affect Climate Far into the Future

Many greenhouse gases stay in the atmosphere for long periods of time. As a result, even if emissions stopped increasing, atmospheric greenhouse gas concentrations would continue to remain elevated for hundreds of years. Moreover, if we stabilized concentrations and the composition of today's atmosphere remained steady (which would require a dramatic reduction in current greenhouse gas emissions), surface air temperatures would continue to warm. This is because the oceans, which store heat, take many decades to fully respond to higher greenhouse gas concentrations. The ocean's response to higher greenhouse gas concentrations and higher temperatures will continue to impact climate over the next several decades to hundreds of years.

Future Temperature Changes

Climate models project the following key temperature-related changes:

Key Global Projections

- Average global temperatures are expected to increase by 2°F to 11.5°F by 2100, depending on the level of future greenhouse gas emissions, and the outcomes from various climate models.
- By 2100, global average temperature is expected to warm at least twice as much as it has during the last 100 years.
- Ground-level air temperatures are expected to continue to warm more rapidly over land than oceans.

- Some parts of the world are projected to see larger temperature increases than the global average.

Future Precipitation and Storm Events

Patterns of precipitation and storm events, including both rain and snowfall are likely to change. However, some of these changes are less certain than the changes associated with temperature. Projections show that future precipitation and storm changes will vary by season and region. Some regions may have less precipitation, some may have more precipitation, and some may have little or no change. The amount of rain falling in heavy precipitation events is likely to increase in most regions, while storm tracks are projected to shift towards the poles. Climate models project the following precipitation and storm changes:

- Global average annual precipitation through the end of the century is expected to increase, although changes in the amount and intensity of precipitation will vary by region.
- The intensity of precipitation events will likely increase on average. This will be particularly pronounced in tropical and high-latitude regions, which are also expected to experience overall increases in precipitation.
- The strength of the winds associated with tropical storms is likely to increase. The amount of precipitation falling in tropical storms is also likely to increase.
- Annual average precipitation is projected to increase in some areas and decrease in others.

Future Ice, Snowpack, and Permafrost

Arctic sea ice is already declining drastically. The area of snow cover in the Northern Hemisphere has decreased since 1970. Permafrost temperature has increased over the last century, making it more susceptible to thawing. Over the next century, it is expected that sea ice will continue to decline, glaciers will continue to shrink, snow cover will continue to decrease, and permafrost will continue to thaw.

For every 2°F of warming, models project about a 15% decrease in the extent of annually averaged sea ice and a 25% decrease in September Arctic sea ice. The coastal sections of the Greenland and Antarctic ice sheets are expected to continue to melt or slide into the ocean. If the rate of this ice melting increases in the 21st century, the ice sheets could add significantly to global sea level rise. Glaciers are expected to continue to decrease in size. The rate of melting is expected to continue to increase, which will contribute to sea level rise.

Future Sea Level Change and Warming temperatures

Warming temperatures contribute to sea level rise by expanding ocean water, melting mountain glaciers and ice caps, and causing portions of the Greenland and Antarctic ice sheets to melt or flow into the ocean. Since 1870, global sea level has risen by about 8 inches. Estimates of future sea level rise vary for different regions, but global sea level for the next century is expected to rise at a greater rate than during the past 50 years. The contribution of thermal expansion, ice caps, and small glaciers to sea level rise is relatively well-studied, but the impacts of climate change on ice sheets are less understood and represent an active area of research. Thus, it is more difficult to predict how much changes in ice sheets will contribute to sea level rise. Greenland and Antarctic ice sheets could contribute an additional 1 foot of sea level rise, depending on how the ice sheets respond.

Regional and local factors will influence future relative sea level rise for specific coastlines around the world. For example, relative sea level rise depends on land elevation changes that occur as a result of subsidence (sinking) or uplift (rising), in addition to things such as local currents, winds, salinity, water temperatures, and proximity to thinning ice sheets. Assuming that these historical geological forces continue, a 2-foot rise in global sea level by 2100 would result in the following relative sea level rise:

- 2.3 feet at New York City
- 2.9 feet at Hampton Roads, Virginia
- 3.5 feet at Galveston, Texas
- 1 foot at Neah Bay in Washington state

Future Ocean Acidification

Ocean acidification is the process of ocean waters decreasing in pH. Oceans become more acidic as carbon dioxide (CO_2) emissions in the atmosphere dissolve in the ocean. This change is measured on the pH scale, with lower values being more acidic. The pH level of the oceans has decreased by approximately 0.1 pH units since pre-industrial times, which is equivalent to a 25% increase in acidity. The pH level of the oceans is projected to decrease even more by the end of the century as CO_2 concentrations are expected to increase for the foreseeable future. Ocean acidification adversely affects many marine species, including plankton, mollusks, shellfish, and corals. As ocean acidification increases, the availability of calcium carbonate will decline. Calcium carbonate is a key building block for the shells and skeletons of many marine organisms. If atmospheric CO_2 concentrations double, coral calcification rates are projected to decline by more than %30. If CO_2 concentrations continue to rise at their current rate, corals could become rare on tropical and subtropical reefs by 2050.

Rise in Global Temperatures and Spread of Disease

This rise in global temperatures will increase the range of disease-carrying insects and the viruses and pathogenic parasites they harbor. Thus, diseases will spread to new regions of the globe. This spread has already been documented with dengue fever, a disease the affects hundreds of millions per year, according to the World Health Organization. Colder temperatures typically limit the distribution of certain species, such as the mosquitoes that transmit malaria, because freezing temperatures destroy their eggs.

Not only will the range of some disease-causing insects expand, the increasing temperatures will also accelerate their lifecycles, which allows them to breed and multiply quicker, and perhaps evolve pesticide resistance faster. In addition to dengue fever, other diseases are expected to spread to new portions of the world as the global climate warms. These include malaria, yellow fever, West Nile virus, zika virus, and chikungunya.

Climate change affects everyone

Our lives are connected to the climate. Human societies have adapted to the relatively stable climate we have enjoyed since the last ice age which ended several thousand years ago. A warming climate will bring changes that can affect our water supplies, agriculture, power and transportation systems, the natural environment, and even our own health and safety.

Carbon dioxide can stay in the atmosphere for nearly a century, on average, so Earth will continue to warm in the coming decades. The warmer it gets, the greater the risk for more severe changes to the climate and Earth's system. Although it's difficult to predict the exact impacts of climate change, what's clear is that the climate we are accustomed to is no longer a reliable guide for what to expect in the future.

We can reduce the risks we will face from climate change. By making choices that reduce greenhouse gas pollution, and preparing for the changes that are already underway, we can reduce risks from climate change. Our decisions today will shape the world our children and grandchildren will live in.

You Can Take Action

You can take steps at home, on the road, and in your office to reduce greenhouse gas emissions and the risks associated with climate change. Many of these steps can save you money. Some, such as walking or biking to work, can even improve your health! You can also get involved on a local or state level to support energy efficiency, clean energy programs, or other climate programs.

AT A GLANCE

Air pollution can be thought of as gaseous and particulate contaminants that are present in the earth's atmosphere. Chemicals discharged into the air that have a direct impact on the environment are called primary pollutants. These primary pollutants sometimes react with other chemicals in the air to produce secondary pollutants. The commonly found air pollutants, known as criteria pollutants, are particle pollution, ground-level ozone, carbon monoxide, sulfur oxides, nitrogen oxides, and lead. These pollutants can harm health and the environment, and cause property damage. The historical record shows that the climate system varies naturally over a wide range of time scales. In general, climate changes prior to the Industrial Revolution in the 1700s can be explained by natural causes, such as changes in solar energy, volcanic eruptions, and natural changes in greenhouse gas concentrations. Recent climate changes, however, cannot be explained by natural causes alone. Natural causes are very unlikely to explain most observed warming, especially warming since the mid-20th century. Rather, human activities can explain most of that warming.

The primary human activity affecting the amount and rate of climate change is greenhouse gas emissions from the burning of fossil fuels. Greenhouse gas concentrations in the atmosphere will continue to increase unless the billions of tons of our annual emissions decrease substantially. Increased concentrations are expected to increase Earth's average temperature, influence the patterns and amounts of precipitation, reduce ice and snow cover, as well as permafrost, raise sea level and increase the acidity of the oceans. These changes will impact our food supply, water resources, infrastructure, ecosystems, and even our own health. Acid rain is a term referring to a mixture of wet and dry deposition from the atmosphere containing higher than normal amounts of nitric and sulfuric acids. The precursors of acid rain formation result from both natural sources, such as volcanoes and decaying vegetation, and man-made sources, primarily emissions of sulfur dioxide (SO_2) and nitrogen oxides (NOx) resulting from fossil fuel combustion. Acid rain causes acidification of lakes and streams, contributes to the damage of trees and many sensitive forest soils. In addition, acid rain accelerates the decay of building materials and paints, contributes to the corrosion of metals and damages human health. The ozone depletion process begins when CFCs and other ozone-depleting substances (ODS) are emitted into the atmosphere. Reductions in stratospheric ozone levels lead to higher levels of UVB reaching the Earth's surface. The sun's output of UVB does not change; rather, less ozone means less protection, and hence more UVB reaches the Earth. Ozone layer depletion increases the amount of UVB tat lead to negative health and environmental effects.

CHAPTER-11

CONCEPTS OF CONVENTIONAL AND SUSTAINABLE ENERGY

11.1 ENERGY USE: CHALLENGES AND IMPACTS

Energy for lighting, heating and cooling our buildings, manufacturing products, and powering our transportation systems comes from a variety of natural sources. The earth's core provides geothermal energy. The gravitational pull of moon and sun create tides. The sun emits light (electromagnetic radiation), which creates wind, powers the water (hydrologic) cycle, and enables photosynthesis. Plants, algae, and cyanobacteria utilize solar energy to grow and create biomass that can be burned and used for **biofuels**, such as wood, biodiesel, bioethanol. Over the course of millions of years, biomass from photosynthetic organisms can create energy-rich **fossil fuels** through the geologic process of burial and transformation through heat and pressure.

Each of these types of energy can be defined as **renewable** or **non-renewable.** Renewable energy sources can be replenished within human lifespans. Examples include solar, wind, and biomass energy. Non-renewable energy is finite and cannot be replenished within a human timescale. Examples include nuclear energy and fossil fuels, which take millions of years to form. All energy sources have and some environmental and health cost, and the distribution of energy is not equally distributed among all nations.

Energy Use: Environmental and Health Challenges

The environmental impacts of energy use on humans and the planet can happen anywhere during the life cycle of the energy source. The impacts begin with the extraction of the resource. They continue with the processing, purification or manufacture of the source; its transportation to place of energy generation, and ends with the disposal of waste generated during use.

Extraction of fossil fuels can be used as a case study because its use has significant impacts on the environment. As we mine deeper into mountains, farther out at sea, or farther into pristine habitats, we risk damaging fragile environments, and the results of accidents or natural disasters during extraction processes can be devastating. Fossils fuels are often located far from where they are utilized so they need to be transported by pipeline, tankers, rail or trucks. These all present the potential for accidents, leakage and spills. When transported by rail or truck energy must be expended and pollutants are generated. Processing of petroleum, gas and coal generates various types of emissions and wastes, as well as utilizes water resources. Production of energy at power plants results in air, water, and, often, waste emissions. Power plants are highly regulated in the Unites States by federal and state law under the Clean Air and Clean Water Acts, while nuclear power plants are regulated by the Nuclear Regulatory Commission.

Geopolitical Challenges of Fossil Fuels

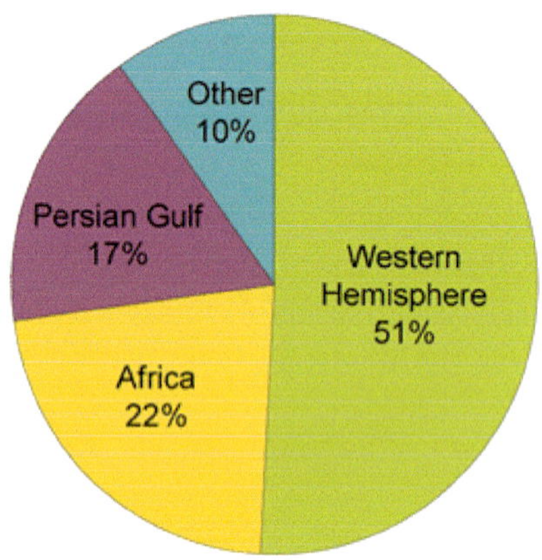

Figure 1. Sources of United States Net Petroleum Imports, 2009 Figure illustrates that the United States imported over half of the crude oil and refined petroleum products that it consumed during 2009. Source: U.S. Energy Information Administration, Petroleum Supply Annual, 2009, preliminary data.

The use of fossil fuels has allowed much of the global population to reach a higher standard of living. However, this dependence on fossil fuels results in many significant impacts on society. Our modern technologies and services, such as transportation and plastics depend in many ways on fossil fuels. If supplies become limited or extremely costly, our economies are vulnerable. If countries do not have fossil fuel reserves of their own, they incur even more risk. The United States has become more and more dependent on foreign oil since 1970 when our own oil production peaked. The United States imported over half of the crude oil and refined petroleum products that we consumed during 2009. Just over half of these imports came from the Western Hemisphere (Figure 2).

The major holder of oil reserves is the Organization of Petroleum Exporting Countries, (OPEC) (Figure 3). As of 2018, there were 15 member countries in OPEC: Algeria, Angola, Congo, Ecuador, Equatorial Guinea, Gabon, Iran, Iraq, Kuwait, Libya, Nigeria, Qatar, Saudi Arabia, the United Arab Emirates, and Venezuela. OPEC attempts to influence the amount of oil available to the world by assigning a production quota to each member except Iraq, for which no quota is presently set.

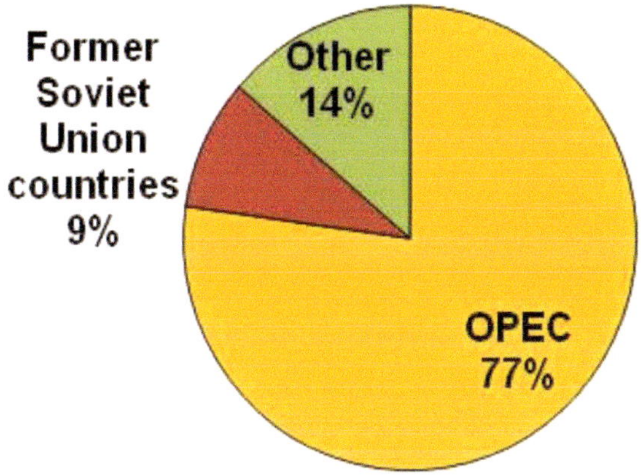

Figure 2. Proven Oil Reserves Holders Pie chart shows proven oil reserves holders. Source: C. Klein-Banai using data from BP Statistical Review of World Energy (2010)

Overall compliance with these quotas is mixed since the individual countries make the actual production decisions. All of these countries have a national oil company but also allow international oil companies to operate within their borders. They can restrict the amounts of production by those oil companies. Therefore, the OPEC countries have a large influence on how much of world demand is met by OPEC and non-OPEC supply. A recent example of this is the price increases that occurred during the year 2011 after multiple popular uprisings in Arab countries, including Libya.

This pressure has lead the United States to developing policies that would reduce reliance on foreign oil such as developing additional domestic sources and obtaining it from non-Middle Eastern countries such as Canada, Mexico, Venezuela, and Nigeria. However, since fossil fuel reserves create jobs and provide dividends to investors, a lot is at stake in a nation that has oil reserves. Oil wealth may be shared with the country's inhabitants or retained by the oil companies and dictatorships, such as in Nigeria prior to the 1990s.

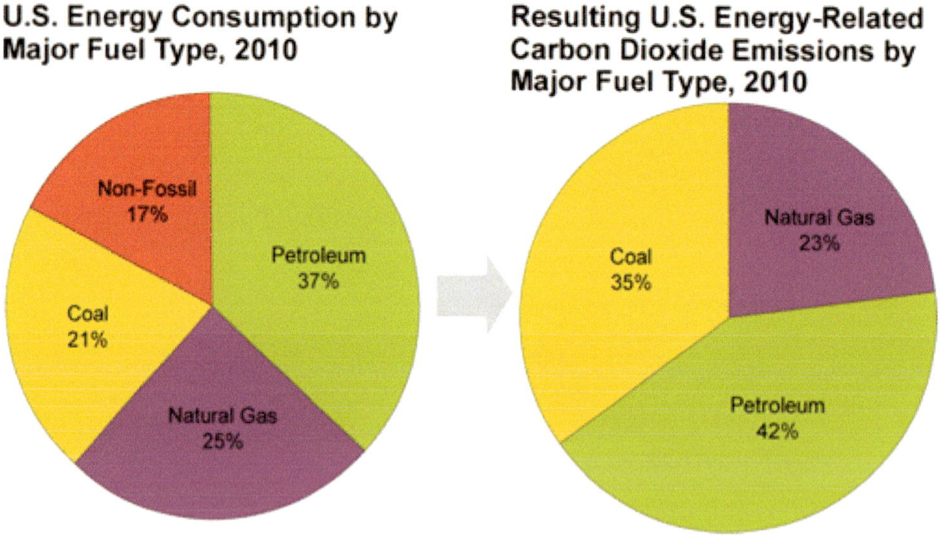

Figure 3. Fuel Type and Carbon Emissions The two charts show the relationship between fuel type and carbon emissions for U.S. energy consumption in 2010. Source: U.S. Energy Information Administration

11.2 NON-RENEWABLE ENERGY RESOURCES

Fossil Fuels

Fossil fuels comes from the organic matter of plants, algae, and cyanobacteria that was buried, heated, and compressed under high pressure over millions of years. The process transformed the biomass of those organisms into the three types of fossil fuels: oil, coal, and natural gas.

Petroleum (oil)

Thirty seven percent of the world's energy consumption and 43% of the United States energy consumption comes from oil. Scientists and policy-makers often discuss the question of when the world will reach peak oil production, the point at which oil production is at its greatest and then declines. It is generally thought that peak oil will be reached by the middle of the 21st Century, although making such estimates is difficult because a lot of variables must be considered. Currently world reserves are 1.3 trillion barrels, or 45 years left at current level of production.

Effects of Oil Extraction and Refining on the Environment

Oil is usually found one to two miles (1.6 – 3.2 km) below the Earth's surface, whether that is on land or ocean. Once oil is found and extracted it must be refined, which separates and prepares the mix of crude oil into the different types for gas, diesel, tar, and asphalt. Oil refining is one of top sources of air pollution in the United States for volatile organic hydrocarbons and toxic emissions, and the single largest source of carcinogenic benzene. When petroleum is burned as gasoline or diesel, or to make electricity or to power boilers for heat, it produces a number of emissions that have a detrimental effect on the environment and human health:

- Carbon dioxide (CO_2) is a greenhouse gas and a source of climate change.
- Sulfur dioxide (SO_2) causes acid rain, which damages plants and animals that live in water, and it increases or causes respiratory illnesses and heart diseases, particularly in vulnerable populations like children and the elderly.
- Nitrous oxides (NO_x) and Volatile Organic Carbons (VOCs) contribute to ozone at ground level, which is an irritant and causes damage to the lungs.
- Particulate Matter (PM) produces hazy conditions in cities and scenic areas, and combines with ozone to contribute to asthma and chronic bronchitis, especially in children and the elderly. Very small, or "fine PM," is also thought to penetrate the respiratory system more deeply and cause emphysema and lung cancer.
- Lead can have severe health impacts, especially for children.

There are other domestic sources of oil that are being considered as conventional resources and are being depleted. These include tar sands – deposits of moist sand and clay with 1-2 percent bitumen (thick and heavy petroleum rich in carbon and poor in hydrogen). These are removed by strip mining (see section below on coal). Another source is oil shale, which is sedimentary rock filled with organic matter that can be processed to produce liquid petroleum. Extracted by strip mining or creating subsurface mines, oil shale can be burned directly like coal or baked in the presence of hydrogen to extract liquid petroleum. However, the net energy values are low and they are expensive to extract and process. Both of these resources have severe environmental impacts due to strip mining, carbon dioxide, methane and other air pollutants similar to other fossil fuels.

As the United States tries to extract more oil from its own dwindling resources, they are drilling even deeper into the earth and increasing the environmental risks. The largest United States oil spill to date began in April 2010 when an explosion occurred on Deepwater Horizon Oil Rig killing 11 employees and spilling nearly 200 million gallons of oil before the resulting leak could be stopped. Wildlife, ecosystems, and people's livelihood were adversely affected. A lot of money and huge amounts of energy were expended on immediate clean-up efforts. The long-term impacts are still not known. The National Commission on the Deepwater Horizon Oil Spill and Offshore Drilling was set up to study what went wrong.

The Dependence of Transportation on Oil on a Global Scale

Two-thirds of oil consumption is devoted to transportation, providing fuel for cars, trucks, trains and airplanes. For the United States and most developed societies, transportation is woven into the fabric of our lives, a necessity as central to daily operations as food or shelter. The concentration of oil reserves in a few regions or the world makes much of the world dependent on imported energy for transportation. The rise in the price of oil in the last decade makes dependence on imported energy for transportation an economic as well as an energy issue. The United States, for example, now spends upwards of $350 billion annually on imported oil, a drain of economic resources that could be used to stimulate growth, create jobs, build infrastructure and promote social advances at home.

Coal

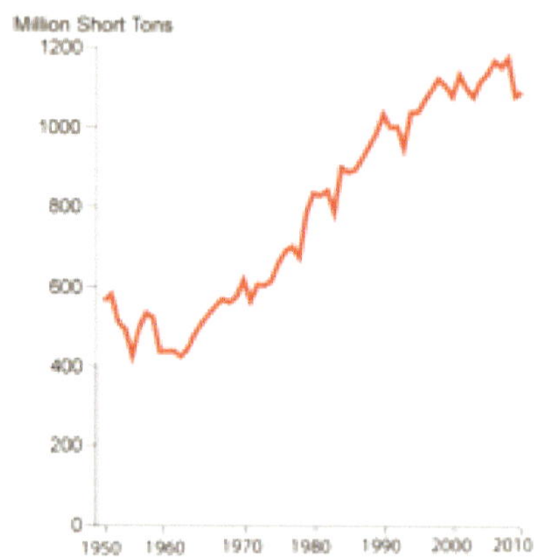

Figure 4. Historic U.S. Coal Production Graph shows U.S. Coal Production from 1950-2010. Source: U.S. Energy Information Administration

Unlike oil, coal is a solid. Due to its relatively low cost and abundance, coal is used to generate about half of the electricity consumed in the United States. Coal is the largest domestically produced source of energy. Coal production has doubled in the United States over the last sixty year (Figure 1). Current world reserves are estimated at 826,000 million tonnes, with nearly 30% of that in the United States. It is a major fuel resource that the United States controls domestically.

Coal is plentiful and inexpensive, when looking only at the market cost relative to the cost of other sources of electricity, but its extraction, transportation, and use produces a multitude of environmental impacts that the market cost does not truly represent. Coal emits sulfur dioxide, nitrogen oxide, and mercury, which have been linked to acid rain, smog, and health issues. Burning of coal emits higher amounts of carbon dioxide per unit of energy than the use of oil or natural gas. Coal accounted for 35% of the total United States emissions of carbon dioxide released into the Earth's atmosphere in 2010. Ash generated from combustion contributes to water contamination. Some coal mining has a negative impact on ecosystems and water quality, and alters landscapes and scenic views (such as with mountaintop mining). There are also significant health effects and risks to coal miners and those living in the vicinity of coal mines. Traditional underground mining is risky to mine workers due to the risk of entrapment or death. Over the last 15 years, the U.S. Mine Safety and Health Administration has published the number of mine worker fatalities and it has varied from 18-48 per year. Twenty-nine miners died on April 6, 2010 in an explosion at the Upper Big Branch coal mine in West Virginia, contributing to the uptick in deaths between 2009 and 2010. In other countries, with less safety regulations, accidents occur more frequently. In May 2011, for example, three people died and 11 were trapped in a coalmine in Mexico for several days. There is also risk of getting black lung disease (pneumoconiosis). This is a disease of the lungs caused by the inhalation of coal dust over a long period of time. It causes coughing and shortness of breath. If exposure is stopped the outcome is good. However, the complicated form may cause shortness of breath that gets increasingly worse.

Figure 5. Mountaintop Removal Coal Mining in Martin County, Kentucky Photograph shows mountaintop coal removal mining in Martin County, Kentucky. Source: Flashdark.

Mountaintop mining (MTM), while less hazardous to workers, has particularly detrimental effects on land resources. MTM is a surface mining practice involving the removal of mountaintops to expose coal seams, and disposing of the associated mining waste in adjacent valleys. This form of mining is very damaging to the environment because it literally removes the tops of mountains, destroying the existing habitat. Additionally, the debris from MTM is dumped into valleys burying streams and other important habitat.

Natural Gas

Natural gas meets 20% of world energy needs and 25% of United States needs. **Natural gas** is mainly composed of methane (CH_4) and is a very potent greenhouse gas. There are two types of natural gas. **Biogenic gas** is found at shallow depths and arises from anaerobic decay of organic matter by bacteria, like landfill gas. **Thermogenic gas** comes from the compression of organic matter and deep heat underground. They are found with petroleum in reservoir rocks and with coal deposits, and these fossil fuels are extracted together.

Natural gas is released into the atmosphere from coal mines, oil and gas wells, and natural gas storage tanks, pipelines, and processing plants. These leaks are the source of about 25% of total U.S. methane emissions, which translates to three percent of total U.S. greenhouse gas emissions. When natural gas is produced but cannot be captured and transported economically, it is "flared," or burned at well sites, which converts it to CO_2. This is considered to be safer and better than releasing methane into the atmosphere because CO_2 is a less potent greenhouse gas than methane. In the last few years a new reserve of natural gas has been identified: shale resources. The United States possesses 2,552 trillion cubic feet (Tcf) (72.27 trillion cubic meters) of potential natural gas resources, with shale resources accounting for 827 Tcf (23.42 tcm). As natural gas prices increased it has become more economical to extract the gas from shale. Figure 3 shows the past and forecasted U.S. natural gas production and the various sources. The current reserves are enough to last about 110 years at the 2009 rate of U.S. consumption (about 22.8 Tcf per year -645.7 bcm per year).

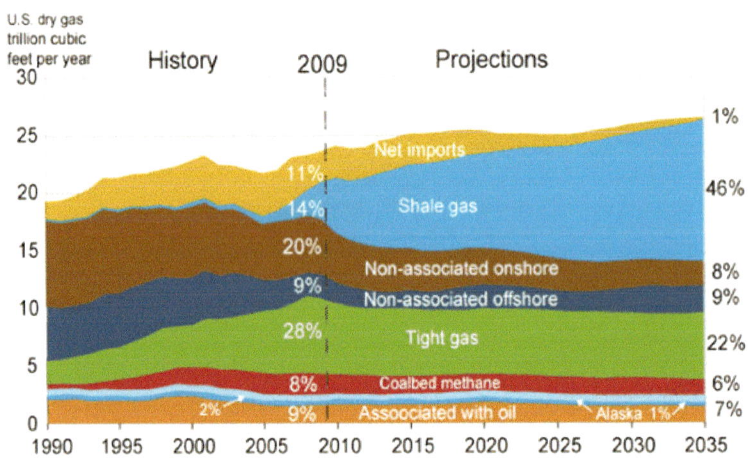

Figure 6. U.S. Natural Gas Supply, 1990-2035 Graph shows U.S. historic and projected natural gas production from various sources. Source: U.S. Energy Information Administration

Natural gas is a preferred fossil fuel when considering its environmental impacts. Specifically, when burned, much less carbon dioxide (CO_2), nitrogen oxides, and sulfur dioxide are omitted than from the combustion of coal or oil. It also does not produce ash or toxic emissions.

Natural gas production can result in the production of large volumes of contaminated water. This water has to be properly handled, stored, and treated so that it does not pollute land and water supplies. Extraction of shale gas is more problematic than traditional sources due to a process nicknamed **fracking,** or fracturing of wells, since it requires large amounts of water (Figure 4). The technique uses high-pressure fluids to fracture the normally hard shale deposits and release gas and oil trapped inside the rock. To promote the flow of gas out of the rock, small particles of solids are included in the fracturing liquids to lodge in the shale cracks and keep them open after the liquids are depressurized. The considerable use of water may affect the availability of water for other uses in some regions and this can affect aquatic habitats. If mismanaged, hydraulic fracturing fluid can be released by spills, leaks, or various other exposure pathways. The fluid contains potentially hazardous chemicals such as hydrochloric acid, glutaraldehyde, petroleum distillate, and ethylene glycol. The risks of fracking have been highlighted in popular culture in the documentary, Gasland (2010).

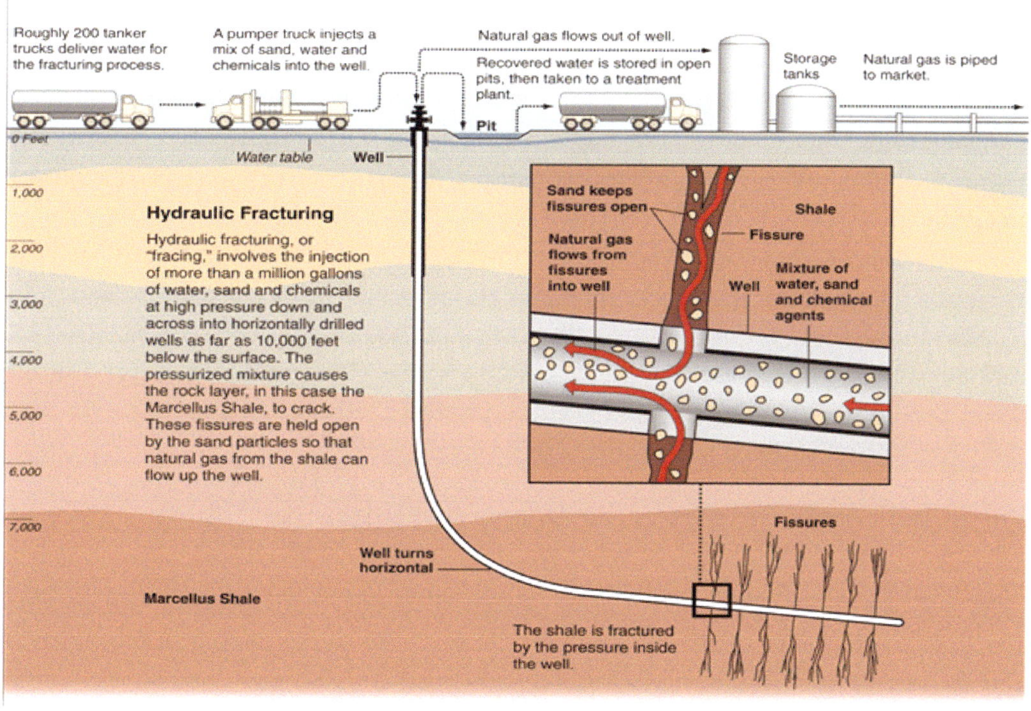

Figure 7. Graphic illustrates the process of hydraulic fracturing. Source: Al Granberg, ProPublica.

The raw gas from a well may contain many other compounds besides the methane that is being sought, including hydrogen sulfide, a very toxic gas. Natural gas with high concentrations of hydrogen sulfide is usually flared which produces CO_2, carbon monoxide, sulfur dioxide, nitrogen oxides, and many other compounds. Natural gas wells and pipelines often have engines to run equipment and compressors, which produce additional air pollutants and noise.

Coal and natural gas contributions to the production of electricity

At present the fossil fuels used for electricity generation in the US are predominantly coal (44%) and natural gas (23%); petroleum accounts for approximately 1%. Coal electricity traces its origins to the early 20th Century, when it was the natural fuel for steam engines given its abundance, high energy density and low cost. gatural Gas is a later addition to the fossil electricity mix, arriving in significant quantities after World War II and with its greatest growth since 1990. Of the two fuels, coal emits almost twice the carbon dioxide as natural gas for the same heat output, making it significantly greater contributor to global warming and climate change.

Future of Coal and Natural Gas

The future development of coal and natural gas depend on the degree of public and regulatory concern for carbon emissions, and the relative price and supply of the two fuels. Supplies of coal are abundant in the United States, and the transportation chain from mines to power plants is well established. The primary unknown factor is the degree of public and regulatory pressure that will be placed on carbon emissions. Strong regulatory pressure on carbon emissions would favor retirement of coal and addition of natural gas power plants. This trend is reinforced by the recent dramatic expansion of shale gas reserves in the United States due to advances in drilling technology. Shale natural gas production has increased 48% annually in the years 2006 – 2010, with more increases expected. Greater United States production of shale gas will gradually reduce imports and could eventually make the United States a net exporter of natural gas.

Nuclear Power: Radioactive Decay of Elements

Nuclear power is energy released from the radioactive decay of elements, such as uranium, which releases large amounts of energy. Nuclear power plants produce no carbon dioxide and, therefore, are often considered an **alternative fuel** (fuels other than fossil fuels). Currently, world production of electricity from nuclear power is about 19.1 trillion KWh, with the United States producing and consuming about 22% of that. Nuclear power provides about 9% of the electricity in the United States (Figure 7).

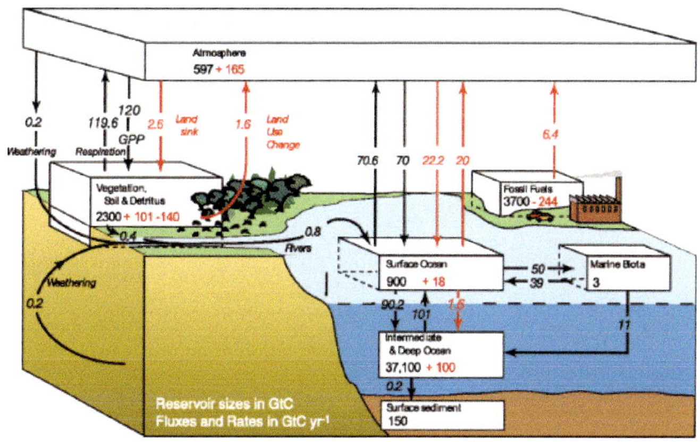

Figure 8. Global Carbon Cycle, 1990s The global carbon cycle for the 1990s, showing the main annual fluxes in GtC yr–1: pre-industrial 'natural' fluxes in black and 'anthropogenic' fluxes in red. Source: Climate Change 2007: The Physical Science Basis: Contribution of Working Group I to the Fourth Assessment Report of the Intergovernmental Panel on Climate Change, Cambridge University Press, figure 7.3

There are environmental challenges with nuclear power. Mining and refining uranium ore and making reactor fuel demands a lot of energy. Also, nuclear power plants are very expensive and require large amounts of metal, concrete, and energy to build. The main environmental challenge for nuclear power is the wastes including uranium mill tailings, spent (used) reactor fuel, and other radioactive wastes. These materials have long radioactive half-lives and thus remain a threat to human health for thousands of years. The **half life** of a radioactive element is the time it takes for 50% of the material to radioactively decay. The U.S. Nuclear Regulatory Commission regulates the operation of nuclear power plants and the handling, transportation, storage, and disposal of radioactive materials to protect human health and the environment.

By volume, the waste produced from mining uranium, called uranium mill tailings, is the largest waste and contains the radioactive element radium, which decays to produce radon, a radioactive gas. High-level radioactive waste consists of used nuclear reactor fuel. This fuel is in a solid form consisting of small fuel pellets in long metal tubes and must be stored and handled with multiple containment, first cooled by water and later in special outdoor concrete or steel containers that are cooled by air. There is no long-term storage facility for this fuel in the United States.

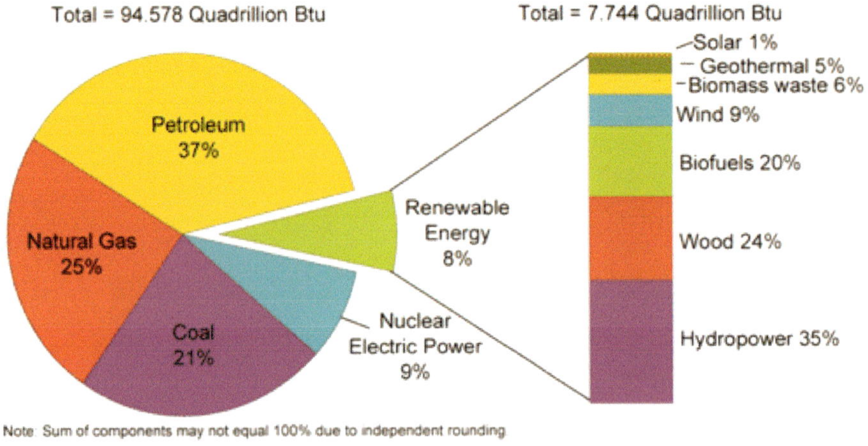

Figure 9. U.S. Energy Consumption by Energy Source, 2009 Renewable energy makes up 8% of U.S. energy consumption. Source: U.S. Energy Information Administration

There are many other regulatory precautions governing permitting, construction, operation, and decommissioning of nuclear power plants due to risks from an uncontrolled nuclear reaction. The potential for contamination of air, water and food is high should an uncontrolled reaction occur. Even when planning for worst-case scenarios, there are always risks of unexpected events. For example, the March 2011 earthquake and subsequent tsunami that hit Japan resulted in reactor meltdowns at the Fukushima Daiichi Nuclear Power Station, causing massive damage to the surrounding area.

Debating Nuclear Energy

From a sustainability perspective, nuclear electricity presents an interesting dilemma. On the one hand, nuclear electricity produces no carbon emissions, a major sustainable advantage in a world facing anthropogenic climate change. On the other hand, nuclear electricity produces dangerous waste that i) must be stored out of the environment for thousands of years, ii) can produce bomb-grade plutonium and uranium that could be diverted by terrorists or others to destroy cities and poison the environment, and iii) threatens the natural and built environment through accidental leaks of long-lived radiation. Thoughtful scientists, policy makers, and citizens must weigh the benefit of this source of carbon-free electricity against the environmental risk of storing spent fuel, the societal risk of nuclear proliferation, and the impact of accidental or deliberate release of radiation. There are very few examples of humans having the power to permanently change the dynamics of the earth. Global climate change from carbon emissions is one example, and radiation from the explosion of a sufficient number of nuclear weapons is another. Nuclear electricity touches both of these opportunities, on the positive side for reducing carbon emissions and on the negative side for the risk of nuclear proliferation.

Nuclear electricity came on the energy scene remarkably quickly. Following the development of nuclear technology at the end of World War II for military ends, nuclear energy quickly acquired a new peacetime path for inexpensive production of electricity. Eleven years after the end of World War II, a very short time in energy terms, the first commercial nuclear reactor produced electricity at Calder Hall in Sellafield, England. The number of nuclear reactors grew steadily to more than 400 by 1990, four years after the Chernobyl disaster in 1986 and eleven years following Three Mile Island in 1979. Since 1990, the number of operating reactors has remained approximately flat, with new construction balancing decommissioning due to public and government reluctance to proceed with nuclear electricity expansion plans.

The outcome of this debate will determine whether the world experiences a nuclear renaissance that has been in the making for several years. The global discussion has been strongly impacted by the unlikely nuclear accident in Fukushima, Japan in March 2011. The Fukushima nuclear disaster was caused by an earthquake and tsunami that disabled the cooling system for a nuclear energy complex consisting of operating nuclear reactors and storage pools for underwater storage of spent nuclear fuel ultimately causing a partial meltdown of some of the reactor cores and release of significant radiation. This event, 25 years after Chernobyl, reminds us that safety and public confidence are especially important in nuclear energy; without them expansion of nuclear energy will not happen.

11.3 RENEWABLE ENERGY RESOURCES

Hydropower

Figure 10. Hoover Power Plant View of Hoover Power Plant on the Colorado River as seen from above. Source: U.S. Department of the Interior

Hydropower (hydroelectric) relies on water to spin turbines and create electricity. It is considered a clean and renewable source of energy because it does not directly produce pollutants and because the source of power is regenerated. Hydropower provides 35% of the United States' renewable energy consumption.

Hydropower dams and the reservoirs they create can have environmental impacts. For example, migration of fish to their upstream spawning areas can be obstructed by dams. In areas where salmon must travel upstream to spawn, such as along the Columbia River in Washington and Oregon, the dams block their way. This problem can be partially alleviated by using "fish ladders" that help salmon get around the dams. Fish traveling downstream, however, can get killed or injured as water moves through turbines in the dam. Reservoirs and operation of dams can also affect aquatic habitats due to changes in water temperatures, water depth, chemistry, flow characteristics, and sediment loads, all of which can lead to significant changes in the ecology and physical characteristics of the river both upstream and downstream. As reservoirs fill with water it may cause natural areas, farms, cities, and archeological sites to be inundated and force populations to relocate.

Small hydropower systems

Large-scale dam hydropower projects are often criticized for their impacts on wildlife habitat, fish migration, and water flow and quality. However, small **run-of- the-river projects** are free from many of the environmental problems associated with their large-scale relatives because they use the natural flow of the river, and thus produce relatively little change in the stream channel and flow. The dams built for some run-of-the-river projects are very small and impound little water, and many projects do not require a dam at all. Thus, effects such as oxygen depletion, increased temperature, decreased flow, and impeded upstream migration are not problems for many run-of-the-river projects.

Small hydropower projects offer emissions-free power solutions for many remote communities throughout the world, such as those in Nepal, India, China, and Peru, as well as for highly industrialized countries like the United States. **Small hydropower systems** are those that generate between .01 to 30 MW of electricity. Hydropower systems that generate up to 100 kilowatts (kW) of electricity are often called **micro hydropower** systems (Figure 2). Most of the systems used by home and small business owners would qualify as microhydropower systems. In fact, a 10 kW system generally can provide enough power for a large home, a small resort, or a hobby farm.

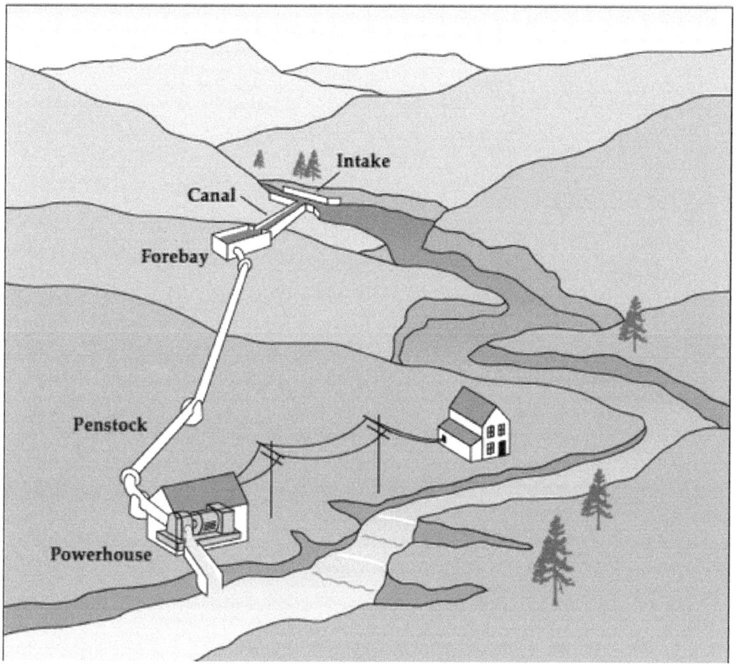

Figure 11. Microhydropower system. Although there are several ways to harness the moving water to produce energy, run-of-the-river systems, which do not require large storage reservoirs, are often used for microhydro, and sometimes for small-scale hydro, projects. For run-of-the-river hydro projects, a portion of a river's water is diverted to a channel, pipeline, or pressurized pipeline (penstock) that delivers it to a waterwheel or turbine. The moving water rotates the wheel or turbine, which spins a shaft. The motion of the shaft can be used for mechanical processes, such as pumping water, or it can be used to power an alternator or generator to generate electricity.

Municipal Solid Waste and Energy

Municipal solid waste (MSW) is commonly known as garbage and can create electricity by burning it directly or by burning the methane produced as it decays. Waste to energy processes are gaining renewed interest as they can solve two problems at once: disposal of waste and production of energy from a renewable resource. Many of the environmental impacts are similar to those of a coal plant: air pollution, ash generation, etc. Because the fuel source is less standardized than coal and hazardous materials may be present in MSW, incinerators and waste-to-energy power plants need to clean the gases of harmful materials. The U.S. EPA regulates these plants very strictly and requires anti-pollution devices to be installed. Also, while incinerating at high temperature many of the toxic chemicals may break down into less harmful compounds. The ash from these plants may contain high concentrations of various metals that were present in the original waste. If ash is clean enough it can be "recycled" as an MSW landfill cover or to build roads, cement block and artificial reefs

Biofuel

Figure 12. Woodchips Photograph shows a pile of woodchips, which are a type of biomass. Source: Ulrichulrich

Biomass refers to material made by organisms, such as cells and tissues. In terms of energy production, biomass is almost always derived from plants, and to a lesser extent, algae. For biomass to be a sustainable option, it usually needs to come from waste material, such as lumber mill sawdust, paper mill sludge, yard waste, or oat hulls from an oatmeal processing plant, material that would otherwise just rot. Livestock manure and human waste could also be considered biomass. The use of biomass can help mitigate climate change because when burned it adds no new carbon to the atmosphere. Thinking back to the carbon cycle (chapter 3), you will recall that photosynthesis removes CO2 through the process of carbon fixation. When biomass is burnt, CO2 is created, but this is equal to the amount of CO2 captured during carbon fixation. Thus, biomass is a carbon neutral energy source because it doesn't add new CO2 to the carbon cycle. Each type of biomass must be evaluated for its environmental and social impact in order to determine if it is really advancing sustainability and reducing environmental impacts. For example, cutting down large swaths of forests just for energy production is not a sustainable option because our energy demands are so great that we would quickly deforest the world, destroying critical habitat.

Burning Wood and Energy

Using wood, and charcoal made from wood, for heating and cooking can replace fossil fuels and may result in lower CO_2 emissions. If wood is harvested from forests or woodlots that have to be thinned or from urban trees that fall down or needed be cut down anyway, then using it for biomass does not impact those ecosystems. However, wood smoke contains harmful pollutants like carbon monoxide and particulate matter. For home heating, it is most efficient and least polluting when using a modern wood stove or fireplace insert that are designed to release small amounts of particulates. However, in places where wood and charcoal are major cooking and heating fuels such as in undeveloped countries, the wood may be harvested faster than trees can grow resulting in deforestation.

Biomass can be used in small power plants. For instance, Colgate College has had a wood-burning boiler since the mid-1980's and in one year it processed approximately 20,000 tons of locally and sustainably harvested wood chips, the equivalent of 1.17 million gallons (4.43 million liters) of fuel oil, avoiding 13,757 tons of emissions and saving the university over $1.8 million in heating costs. The University's steam-generating wood-burning facility now satisfies more than 75% of the campus's heat and domestic hot water needs.

Landfill Gas or Biogas Fuels

Landfill gas (**biogas**) is a sort of man-made "biogenic" gas as discussed above. Methane is formed as a result of biological processes in sewage treatment plants, waste landfills, anaerobic composting, and livestock manure management systems. This gas is captured and burned to produce heat or electricity. The electricity may replace electricity produced by burning fossil fuels and result in a net reduction in CO_2 emissions. The only environmental impacts are from the construction of the plant itself, similar to that of a natural gas plant.

Bioethanol and Biodiesel

Bioethanol and biodiesel are liquid biofuels manufactured from plants, typically crops. Bioethanol can be easily fermented from sugar cane juice, as is done in Brazil. Bioethanol can also be fermented from broken down corn starch, as is mainly done in the United States. The economic and social effects of growing plants for fuels need to be considered, since the land, fertilizers, and energy used to grow biofuel crops could be used to grow food crops instead. The competition of land for fuel vs. food can increase the price of food, which has a negative effect on society. It could also decrease the food supply increasing malnutrition and starvation globally. Also, in some parts of the world, large areas of natural vegetation and forests have been cut down to grow sugar cane for bioethanol and soybeans and palm-oil trees to make biodiesel. This is not sustainable land use. Biofuels may be derived from parts of plants not used for food, such as stalks, thus reducing that impact. Biodiesel can be made from used vegetable oil and has been produced on a very local basis. Compared to petroleum diesel, biodiesel combustion produces less sulfur oxides, particulate matter, carbon monoxide, and unburned and other hydrocarbons, but it produces more nitrogen oxide.

Liquid biofuels typically replace petroleum and are used to power vehicles. Although ethanol-gasoline mixtures burn cleaner than pure gasoline, they also are more volatile and thus have higher "evaporative emissions" from fuel tanks and dispensing equipment. These emissions contribute to the formation of harmful, ground level ozone and smog. Gasoline requires extra processing to reduce evaporative emissions before it is blended with ethanol.

Geothermal Energy: Endless Energy

Five percent of the United States' renewable energy comes from geothermal energy: using the heat of Earth's subsurface to provide endless energy. Geothermal systems utilize a heat-exchange system that runs in the subsurface about 20 feet (5 meters) below the surface where the ground is at a constant temperature. The system uses the earth as a heat source (in the winter) or a heat sink (in the summer). This reduces the energy consumption required to generate heat from gas, steam, hot water, and conventional electric air-conditioning systems.The environmental impact of geothermal energy depends on how it is being used. Direct use and heating applications have almost no negative impact on the environment.

Geothermal power plants do not burn fuel to generate electricity so their emission levels are very low. They release less than 1% of the carbon dioxide emissions of a fossil fuel plant. Geothermal plants use scrubber systems to clean the air of hydrogen sulfide that is naturally found in the steam and hot water. They emit 97% less acid rain-causing sulfur compounds than are emitted by fossil fuel plants. After the steam and water from a geothermal reservoir have been used, they are injected back into the earth.

Solar Energy

Figure 13. Rooftop Solar Installations Rooftop solar installation on Douglas Hall at the University of Illinois at Chicago has no effect on land resources, while producing electricity with zero emissions. Source: Office of Sustainability, UIC

Solar power converts the energy of light into electrical energy and has minimal impact on the environment, depending on where it is placed. In 2009, 1% of the renewable energy generated in the United States was from solar power (1646 MW) out of the 8% of the total electricity generation that was from renewable sources. The manufacturing of photovoltaic (PV) cells generates some hazardous waste from the chemicals and solvents used in processing. Often solar arrays are placed on roofs of buildings or over parking lots or integrated into construction in other ways. However, large systems may be placed on land and particularly in deserts where those fragile ecosystems could be damaged if care is not taken. Some solar thermal systems use potentially hazardous fluids (to transfer heat) that require proper handling and disposal. Concentrated solar systems may need to be cleaned regularly with water, which is also needed for cooling the turbine-generator. Using water from underground wells may affect the ecosystem in some arid locations.

Wind Energy

Wind energy is a renewable energy source that is clean and has very few environmental challenges. Wind turbines are becoming a more prominent sight across the United States, even in regions that are considered to have less wind potential. Wind turbines (often called windmills) do not release emissions that pollute the air or water (with rare exceptions), and they do not require water for cooling. The U.S. wind industry had 40,181 MW of wind power capacity installed at the end of 2010, with 5,116 MW installed in 2010 alone, providing more than 20% of installed wind power around the globe. According to the American Wind Energy Association, over 35% of all new electrical generating capacity in the United States since 2006 was due to wind, surpassed only by natural gas.

Figure 14. Twin Groves Wind Farm, Illinois Wind power is becoming a more popular source of energy in the United States. Source: Office of Sustainability, UIC

Because a wind turbine has a small physical footprint relative to the amount of electricity it produces, many wind farms are located on crop and pasture land. They contribute to economic sustainability by providing extra income to farmers and ranchers, allowing them to stay in business and keep their property from being developed for other uses. For example, energy can be produced by installing wind turbines in the Appalachian mountains of the United States instead of engaging in mountain top removal for coal mining. Offshore wind turbines on lakes or the ocean may have smaller environmental impacts than turbines on land.

Wind turbines do have a few environmental challenges. There are aesthetic concerns to some people when they see them on the landscape. A few wind turbines have caught on fire, and some have leaked lubricating fluids, though this is relatively rare. Some people do not like the sound that wind turbine blades make. Turbines have been found to cause bird and bat deaths particularly if they are located along their migratory path. This is of particular concern if these are threatened or endangered species. There are ways to mitigate that impact and it is currently being researched. There are some small impacts from the construction of wind projects or farms, such as the construction of service roads, the production of the turbines themselves, and the concrete for the foundations. However, overall analysis has found that turbines make much more energy than the amount used to make and install them.

Interest in Renewable Energy

Strong interest in renewable energy in the modern era arose in response to the oil shocks of the 1970s, when the Organization of Petroleum Exporting Countries (OPEC) imposed oil embargos and raised prices in pursuit of geopolitical objectives. The shortages of oil, especially gasoline for transportation, and the eventual rise in the price of oil by a factor of approximately 10 from 1973 to 1981 disrupted the social and economic operation of many developed countries and emphasized their precarious dependence on foreign energy supplies. The reaction in the United States was a shift away from oil and gas to plentiful domestic coal for electricity production and the imposition of fuel economy standards for vehicles to reduce consumption of oil for transportation. Other developed countries without large fossil reserves, such as France and Japan, chose to emphasize nuclear (France to the 80% level and Japan to 30%) or to develop domestic renewable resources such as hydropower and wind (Scandinavia), geothermal (Iceland), solar, biomass and for electricity and heat. As oil prices collapsed in the late 1980s interest in renewables, such as wind and solar that faced significant technical and cost barriers, declined in many countries, while other renewables, such as hydropower and biomass, continued to experience growth.

The increasing price and volatility of oil prices since 1998, and the increasing dependence of many developed countries on foreign oil (60% of United States and 97% of Japanese oil was imported in 2008) spurred renewed interest in renewable alternatives to ensure energy security. A new concern, not known in previous oil crises, added further motivation: our knowledge of the emission of greenhouse gases and their growing contribution to climate change. An additional economic motivation, the high cost of foreign oil payments to supplier countries (approximately $350 billion/year for the United States at 2011 prices), grew increasingly important as developed countries struggled to recover from the economic recession of 2008. These energy security, carbon emission, and climate change concerns drive significant increases in fuel economy standards, fuel switching of transportation from uncertain and volatile foreign oil to domestic electricity and biofuels, and production of electricity from low carbon sources.

Physical Origin of Renewable Energy

Although renewable energy is often classified as hydro, solar, wind, biomass, geothermal, wave and tide, all forms of renewable energy arise from only three sources: the light of the sun, the heat of the earth's crust, and the gravitational attraction of the moon and sun. Sunlight provides by far the largest contribution to renewable energy. The sun provides the heat that drives the weather, including the formation of high- and low-pressure areas in the atmosphere that make wind. The sun also generates the heat required for vaporization of ocean water that ultimately falls over land creating rivers that drive hydropower, and the sun is the energy source for photosynthesis, which creates biomass. Solar energy can be directly captured for water and space heating, for driving conventional turbines that generate electricity, and as excitation energy for electrons in semiconductors that drive photovoltaics. The sun is also responsible for the energy of fossil fuels, created from the organic remains of plants and sea organisms compressed and heated in the absence of oxygen in the earth's crust for tens to hundreds of millions of years. The time scale for fossil fuel regeneration, however, is too long to consider them renewable in human terms.

Geothermal energy originates from heat rising to the surface from earth's molten iron core created during the formation and compression of the early earth as well as from heat produced continuously by radioactive decay of uranium, thorium and potassium in the earth's crust. Tidal energy arises from the gravitational attraction of the moon and the more distant sun on the earth's oceans, combined with rotation of the earth. These three sources – sunlight, the heat trapped in earth's core and continuously generated in its crust, and gravitational force of the moon and sun on the oceans – account for all renewable energy.

Capacity and Geographical Distribution

Although renewable energies such as wind and solar have experienced strong growth in recent years, they still make up a small fraction of the world's total energy needs. The largest share comes from traditional biomass, mostly fuel wood gathered in traditional societies for household cooking and heating, often without regard for sustainable replacement. Hydropower is the next largest contributor, an established technology that experienced significant growth in the 20th Century. The other contributors are more recent and smaller in contribution: water and space heating by biomass combustion or harvesting solar and geothermal heat, biofuels derived from corn or sugar cane, and electricity generated from wind, solar and geothermal energy. Wind and solar electricity, despite their large capacity and significant recent growth, still contributed less than 1% of total energy in 2008.

The potential of renewable energy resources varies dramatically. Solar energy is by far the most plentiful, delivered to the surface of the earth at a rate of 120,000 Terawatts (TW), compared to the global human use of 15 TW. To put this in perspective, covering 100×100 km2 of desert with 10% efficient solar cells would produce 0.29 TW of power, about 12% of the global human demand for electricity. To supply all of the earth's electricity needs (2.4 TW in 2007) would require 7.5 such squares, an area about the size of Panama (0.05% of the earth's total land area). The world's conventional oil reserves are estimated at three trillion barrels, including all the oil that has already been recovered and that remain for future recovery. The solar energy equivalent of these oil reserves is delivered to the earth by the sun in 1.5 days.

The geographical distribution of useable renewable energy is quite uneven. Sunlight, often thought to be relatively evenly distributed, is concentrated in deserts where cloud cover is rare. Winds are up to 50% stronger and steadier offshore than on land. Hydroelectric potential is concentrated in mountainous regions with high rainfall and snowmelt. Biomass requires available land that does not compete with food production, and adequate sun and rain to support growth.

AT A GLANCE

We derive our energy from a multitude of resources that have varying environmental challenges related to air and water pollution, land use, carbon dioxide emissions, resource extraction and supply, as well as related safety and health issues. Each resource needs to be evaluated within the sustainability paradigm. Coal (45 percent) and gas (23 percent) are the two primary fossil fuels for electricity production in the United States. Coal combustion produces nearly twice the carbon emissions of gas combustion. Increasing public opinion and regulatory pressure to lower carbon emissions are shifting electricity generation toward gas and away from coal. Oil for transportation and electricity generation are the two biggest users of primary energy and producers of carbon emissions in the United States. Transportation is almost completely dependent on oil and internal combustion engines for its energy. The concentration of oil in a few regions of the world creates a transportation energy security issue. Nuclear electricity offers the sustainable benefit of low carbon electricity at the cost of storing spent fuel out of the environment for up to hundreds of thousands of years. Reprocessing spent fuel offers the advantages of higher energy efficiency and reduced spent fuel storage requirements with the disadvantage of higher risk of weapons proliferation through diversion of the reprocessed fuel stream.

Strong interest in renewable energy arose in the 1970s as a response to the shortage and high price of imported oil, which disrupted the orderly operation of the economies and societies of many developed countries. Today there are new motivations, including the realization that growing greenhouse gas emission accelerates global warming and threatens climate change, the growing dependence of many countries on foreign oil, and the economic drain of foreign oil payments that slow economic growth and job creation. There are three ultimate sources of all renewable and fossil energies: sunlight, the heat in the earth's core and crust, and the gravitational pull of the moon and sun on the oceans. Renewable energies are relatively recently developed and typically operate at lower efficiencies than mature fossil technologies. Like early fossil technologies, however, renewables can be expected to improve their efficiency and lower their cost over time, promoting their economic competitiveness and widespread deployment. The future deployment of renewable energies depends on many factors, including the availability of suitable land, the technological cost of conversion to electricity or other uses, the costs of competing energy technologies, and the future need for energy.

Glossary

ABIOTIC: nonliving components of an ecosystem.

ACCLIMATION: to become accustomed to a new climate or environment; to physically adapt.

ACTIVE INGREDIENT: an ingredient that is biologically active in a formulated product.

AGENT ORANGE: an herbicide used by the U.S. military to defoliate forests during the Vietnam War.

AGRO-ECOSYSTEM: The area of land that is influenced by or that supports agricultural activity; land used for crops, pasture, and livestock, the adjacent uncultivated land that supports other vegetation and wildlife and the associated atmosphere, the underlying soil, ground and water, irrigation channels, and drainage networks.

AIR QUALITY INDEX (AQI): **a numerical index used by government agencies to** inform the public of air quality levels.

ALGAL BLOOM: a rapid increase of accumulation in the population of algae (typically microscopic) in an aquatic system.

ALKYLPHENOL: a commonly accepted marker of industrial and urban pollution; a family of organic compounds used in the production of detergents and other cleaning products.

ALLUVIUM: a deposit of sand, mud, or other sediment formed by flowing water.

ALPHA (MALE/FEMALE): the highest ranking member of a group of animals that has reserved breeding rights and first choice of food from a kill.

ANADROMOUS: a species which migrates from the sea to freshwater to reproduce.

ANNUAL CROPS: crops that grow, produce seeds, and die in a year and must be replanted each season.

ANTHROPOGENIC: caused by or related to human action; originating in human activity.

ANTI-SCALANT: a chemical that inhibits or delays precipitation and subsequent scale formation.

AQUACULTURE: Farming of aquatic organisms (plant or animal) in any water environment (ocean, pond, or river).

AQUEDUCT SYSTEM: a series of pipes, ditches, canals, tunnels, and supporting structures used to transport water from its source to its main distribution point.

AQUIFER: an underground layer of permeable rock or soil that is saturated with water; a geological formation containing ground water.

ARID: a region characterized by a severe lack of available water to the extent of hindering or preventing the growth and development of plant and animal life.

ATMOSPHERE: the layer of gases surrounding Earth and held by Earth's gravity.

ATMOSPHERIC RIVERS: regions of intense moisture over oceans that transport moisture from lower altitudes to higher altitudes.

ATOLL: a region characterized by a severe lack of available water to the extent of hindering or preventing the growth and development of plant and animal life.

BACTERIA: a single-celled, prokaryotic organisms that lacks a nuclei.

BALLAST WATER: water taken into the haul of a shipping vessel to provide stability.

BASHAR AL-ASSAD: the President of Syria since 2000.

BIODIVERSITY: the variety of life on Earth or in a particular habitat or ecosystem.

BIOLOGICAL CONTROL: biocontrol, a management practice that involves introducing a predator to an area to regulate the population of a pest.

BIOMAGNIFICATION: the increased level of substances within the tissue of predators as a result of consuming organisms that contain bio-accumulated substances such as mercury.

BIOMASS: material from plants and animals that can be used for energy.

BIOTIC: the living or organic components of an ecosystem.

BIRTH DEFECT: structural or functional abnormalities present at birth that cause physical or mental disability.

BORNEO: island in Indonesia, east of Sumatra, which is mountainous and has extensive rainforests.

BRINE: water saturated with salt; a liquid with high salinity.

BROADCAST: to uniformly distribute fertilizer on a soil surface without working it into the soil.

BYCATCH: a non-target organism that is caught unintentionally when harvesting a target species.

CALCIUM CARBONATE: caco3; a common chemical compound composed of three main elements calcium, carbon and oxygen.

CANCER: a type of disease characterized by unregulated cell growth.

CANINE DISTEMPER: a viral disease associated with canines that affects the lymphatic and nervous system.

CARBON CREDITS: a permit that represents carbon removed from the atmosphere, which can be purchased by companies or governments to offset the carbon emissions they generate.

CARBON SINK: a process or material in an environment that removes carbon dioxide from the atmosphere.

CARCINOGENIC: cancer causing.

CARRYING CAPACITY: the population size that an area can support indefinitely.

CERES: a non-profit organization advocating for sustainability leadership.

CESIUM-137: one of the common fission products of uranium nuclear power plants.

CLEAN AIR ACT: U.S. law that authorizes the EPA to set standards for dangerous air pollutants and enforce those standards.

CLEAN WATER ACT: U.S. federal legislation that regulates the release of point source pollution into surface waters and sets water quality standards for those waters.

CLIMATE: long-term weather conditions in a particular region.

CLIMATE CHANGE: a change in global climate patterns such as rainfall and average regional temperature.

CLIMATOLOGY: the scientific study of Earth's climates, climate variability, climate change, and effects of climate on the biosphere.

CNIDARIAN: an invertebrate belonging to the Cnidarian phylum; they are characterized to have radial symmetry and a saclike internal cavity.

COLIFORM BACTERIA: type of bacteria that are universally found in the feces of warm blooded animals; commonly used as a water quality indicator.

COLONIES: refers to individual organisms of the same species living closely together.

COLONY COLLAPSE DISORDER (CCD): a disorder in honey bees categorized by sudden loss of worker bees, lack of dead bodies, and delayed invasion from attackers.

COMMODITY: a physical substance that is uniform in quality and can be bought and sold.

CONDENSE: to change from a gas to a liquid.

CONDUCTOR: a material which easily permits the flow of electrons, often an electric current.

CONSERVATION: a political, environmental, and social movement to help protect natural resources and wildlife from exploitation.

CONTAMINATION: to make something dangerous, dirty or impure by adding something harmful or undesirable to it

CONTRACEPTION: a technique used to prevent pregnancy

CORAL BLEACHING: the expulsion of the algae living within coral's tissues causing it to die

CORRELATION: a statistical relationship involving the dependence of two or more variables

CRITICALLY ENDANGERED: a species that is facing a very high risk of extinction

CRITTERCAM: a camera that is attached to a wild animal with intent to study the animal's behavior and ecology

CROP YIELD: the amount of product a crop produces per unit of land in a given time

CULLING: a method of population control that involves the killing of an entire family unit

DAMASCUS: the capital city of Syria

DEFORESTATION: removal of forest areas by humans for non-forest use, often for agriculture

DESALINATION: the removal of salt from a water source to convert it to fresh water, often used with seawater

DESERTIFICATION: the increased degradation of drought affected areas, causing them to become deserts

DIABETES: a metabolic disorder defined by high blood sugar and insulin resistance

DIOXINS: a group of chemically related compounds that are classified as environmental pollutants

DISRUPTIVE STIMULI: chemical, light, or audio repellents used to try and deter organisms

DISTILLATION: the process of separating substances in a mixture

DNA: deoxyribonucleic acid; found in all living organisms and contains genetic information

DROUGHT: an extended amount of time with little to no precipitation

ECOLOGICAL SOCIETY OF AMERICA: a nonpartisan, nonprofit organization of scientists founded to promote, conduct, and disseminate studies in the ecological sciences

ECOSYSTEM: all of the organisms in a given area interacting in a physical environment

EDNA: Environmental DNA; genetic material left behind by a species

EFFLUENT: liquid waste discharged into a body a water; outflow from a sewage treatment plant that is rich in nutrients

ELECTRIC FISH BARRIER: a tool that sends low-level electric currents into a body of water to cause discomfort in fish and cause them move away from the affected area; used to regulate the movement of fish

ELYTRA: one of the anterior wings in beetles and other insects that serve to protect the delicate posterior pair of functional wings

EMERALD ASH BORER: a green beetle that was accidentally transported from Asia to North America in the 1990s and has caused widespread destruction of ash trees in the United States and Canada

EMISSIONS: the production and discharge of a gas

ENDANGERED: a species that is threatened with extinction

ENDOCRINE DISRUPTOR: a chemical that interferes with the endocrine system, typically mimicking a hormone or preventing a hormone from having effect

ENDOCRINE SYSTEM: the system cells, glands, and tissue within an organism that is responsible for secreting hormones into the bloodstream

ENTRAINMENT: the loss of fish during water diversion often occurring when fish enter irrigation systems and become isolated when diversion ends

ENVIRONMENTAL PROTECTION AGENCY (EPA): a federal agency in the United States tasked with implementing legislation concerning the protection of the environment

ENVIRONMENTAL SUSTAINABILITY INDEX: a measure of a country's overall progress towards environmental sustainability

ENZYME: a molecular protein that acts as a catalysts and facilitates complex biological reactions

EROSION: the process by which something is gradually worn away by natural forces (wind, water, or other natural agents)

EUROPEAN UNION: economic and political union of 28 member states that are located primarily in Europe

EUTROPHICATION: the process by which a body of water is enriched with nutrients (nitrogen and phosphorus) stimulating excessive growth of photosynthetic organisms like algae. Human activities can accelerate the process.

EXTINCTION: the complete loss of a species

EXTIRPATION: describes a species that is locally extinct in one or more areas within its historical range

FALLOUT: airborne radioactive particles that gradually fall back to Earth after a nuclear blast

FARADAY'S LAW: Faraday's Law of induction states that the flow of electricity will be induced on a conductor when it is in the presence of a varying magnetic field. This principle allows for the generation of electricity by moving magnets through coils of wire.

FECUNDITY: related to fertility; the actual reproductive rate of an organism

FIELD TRIALS: small, controlled fields of genetically modified plants used to conduct research

FLOY TAG: a small piece of plastic attached to live fish as part of scientific studies; a company which produces tags used to specifically identify live fish as part of studies

FOLSOM LAKE: a reservoir in Northern California formed by the Folsom Dam and that provides flood control, drinking water, hydroelectricity, and water for irrigation to the surrounding communities

FOOD AND DRUG ADMINISTRATION (FDA): a federal agency of the United States government found within the Department of Health and Human Services which is responsible for protecting the public health by assuring the safety, efficacy and security of human and veterinary drugs, biological products, medical devices, food supply, cosmetics, and products that emit radiation

FOSTERING: when a female cares for and raises an orphaned juvenile in addition to her own offspring or after the death of her biological offspring

FOUL BROOD DISEASE: a disease in bees in which bacteria attack larvae and kill them before they mature

FRACKING FLUID: a combination of sand, liquids, and chemicals used in the hydraulic fracturing process which are pumped underground to break away natural gasses from permeable rock layers

FUNCTIONAL EXTINCTION: when a population becomes so reduced that it no longer plays an important role in the way an ecosystem functions

FUNGUS: a diverse group of eukaryotic single-celled or multicelled organisms that live by decomposing organic material

GAME RESERVATION: a controlled area used for wildlife conservation which sometimes allows regulated hunting

GENE: the hereditary material of an organism that direct the production of a particular protein and influence an individual's traits

GENETIC DIVERSITY: the hereditary variability among individuals of a single population or within a species

GENETIC MARKER: a gene or short sequence of DNA used to identify a chromosome or to locate other genes on a genetic map

GENETIC MUTATION: a heritable change in the DNA sequence

GENOTYPE: the complete genetic content of an organism

GHOST TOWN: a human settlement, such as a town, that has few or no remaining inhabitants

GLOBAL WARMING: a scientifically observed and ongoing rise in the biosphere's average temperature that is contributing to climate change

GMO: a genetically modified organism; an organism whose DNA has been altered with a gene from another species to produce a desired trait

GOLDEN RICE: a variety of rice that has been genetically modified to produce beta-carotene, a precursor of vitamin A

GRAPHITE MODERATOR: a nuclear reactor that uses carbon as a neutron moderator, which allows un.enriched uranium to be used as nuclear fuel

GREENHOUSE GASES: water vapor, carbon dioxide, methane, nitrous oxide, and ozone; gases that contribute the greenhouse effect and global climate change

GROSS DOMESTIC PRODUCT (GDP): the total value of goods and services provided in a country during a one year period

GROUNDWATER: the water beneath the surface of the ground, consisting largely of surface water that seeped into the ground

HABITAT: the environment in which an animal or plant species resides

HALLIBURTON LOOPHOLE: a provision included in the U.S. Energy Policy Act of 2005 that gave natural gas drilling and extraction companies exemptions from the Safe Drinking Water Act

HAULING OUT: leaving the water to stay on land between periods of foraging

HEGEMONY: a form of government where one state or group has cultural, economic, and military dominance over all others

HERBICIDE: a substance used to exterminate or slow the growth of plants

HIBERNATION: cessation from or slowing of activity during the winter

HIGH PRESSURE RIDGE: a region where the atmosphere pressure at the surface of the planet if greater than its surrounding environment

HISTOPATHOLOGY: the study of tissue abnormalities using microscopy

HOME RANGE: the area where an organism lives and travels

HYDRAULIC FRACTURING: aka fracking; the process of breaking up subsurface rock formations through the use of high pressure liquids usually mixed with sand and chemicals to extract gas and oil

HYDROGEOLOGY: the study of the distribution and movement of water through the Earth

HYDROGRAPH: a graph showing the rate by which water flows past a specific point in a river over time, or a chart that displays a hydrologic variable over time

HYPOCHLORITE: any salt or ester of hydroclorus acid; a chemical commonly found in bleach, water purifiers, and cleaning products

HYPOXIA: a condition in which a body of water contains inadequate amount of oxygen, compromising the health of aquatic organisms

IMMUNOCONTRACEPTION: a type of contraceptive technique that uses immunochemistry as opposed to hormones to prevent pregnancy

INACTIVE INGREDIENT: an ingredient that does not increase or affect the intended action or purpose of a formulated product, drug, or pesticide

INFRASTRUCTURE: the physical and organizational structures and facilities needed for a functioning society

INHOSPITABLE: an environment where organisms cannot grow or live easily

INSECTICIDE: chemicals used to kill insects

INVASIVE SPECIES: a nonnative species whose introduction is likely to cause environmental or economic harm; a nonnative species that disrupts the local ecosystem

IODINE-131: heavy radioactive isotope of iodine with a half-life of 8 days; a major product produced by uranium plutonium nuclear power plants

ISOTOPE: an atom of a given element that differs in the number of neutrons in its nucleus

IVORY: mammalian teeth or tusk composed of mostly dentin, the desirable white material in which elephant tusks are made

KEYSTONE SPECIES: a species whose presence in its ecosystem is necessary to prevent the system from collapse

LACTATION: when mammary glands (breast tissue) secretes milk, often occurring after a female has given birth and is supporting offspring

LANDSAT: a satellite imagery program first launched in 1972

LARVAE: the immature and often wormlike feeding form that hatches from the egg of many insects that have a metaphoric lifecycle

LETHAL CONTROL: a method of controlling wildlife populations that result in the death of individuals from the target population

LIMITING NUTRIENT: a nutrient in limited supply relative to others that when exhausted prevents further growth of an organism

MACRONUTRIENT: in the case of plants, phosphorus, nitrogen, and potassium; a nutrient needed in large amounts for normal growth to occur

MATRIARCH: the female who is the leader of a family unit

MICROORGANISM: an organism that cannot be seen with the human eye

MICROSCOPY: the practice of using a microscope to study objects

MIDDLE EAST: geographic region northeast of Africa and southwest of Asia that contains the country Syria

MIOCENE AGE: the first geological epoch of the Neogene period and extends from about 23.03 to 5.332 million years ago, this time period had a warmer global climate

MISSISSIPPI RIVER BASIN: the area covering much of the central United States that includes tributaries that drain into the Mississippi River

MOBBING: when more than one adult male attempts to mount a female in heat; often leads to serious injury and has been linked to a boost in testosterone and uneven distribution of gender in a population

MODIS IMAGES: Moderate Resolution Imaging Spectroradiometer; a satellite imaging instrument that views the entire Earth's surface every 1-2 days

MONOCROP: a single crop produced year after year on the same land that is economically efficient but ultimately damaging to soil ecology

MONSANTO: an American multinational chemical and agricultural biotechnology company; the leading producer of genetically engineered seeds

MUSTH: a period in which a bull (male) elephant is ready to mate, characterized by increased aggression and a surge of reproductive hormones – mainly testosterone

NATIONAL ACADEMY OF SCIENCES: a non-profit society established by the U.S. congress in 1863 in order to provide independent and objective consultation on matters related to science and technology

NATIONAL PARKS: an area of land set aside by the national government of a country specifically for the purpose of safeguarding nature and wildlife biodiversity

NATIONAL TIGER RECOVERY PROGRAM: a conservation program enacted in 2011 by the Ministry of Forestry in Indonesia with the goal of doubling the number of wild Sumatran tigers by 2022

NATURAL GAS: a combustible mixture of gaseous hydrocarbons consisting mostly of methane

NEBRASKA SANDHILLS: a region of mixed-grass prairie on grass-stabilized sand dunes, it is the largest sand dune formation in the western hemisphere

NO-TILL: growing crops without disturbing the soil through tilling or plowing in order to increase soil organic matter and decrease soil erosion

NOBLE GAS: any of the gaseous elements helium, neon, argon, krypton, xenon, and radon, occupying Group 0 (18) of the periodic table that are believed to be unreactive

NON-GMO PROJECT: a non-profit organization that verifies non-GMO products and educates consumers

NON-LETHAL CONTROL: a method of controlling wildlife populations through actions that do not harm the health of the target population

NUCLEOTIDE SEQUENCE: the order in which nucleotides are situated in a chain relative to one another, which provide the template of a particular amino acid

OCEAN ACIDIFICATION: the decrease of ph levels in the ocean due to the uptake of carbon dioxide from the atmosphere

OGALLALA AQUIFER: an aquifer located on the High Plains in the United States, it is composed of clay, sand, silt, and gravel

OZONE LAYER: a region of Earth's atmosphere that absorbs most of the Sun's ultraviolet (UV) radiation

PAPAYA RINGSPOT VIRUS (PRSV); a virus that infects papayas and is transmitted by aphids. Infested papayas exhibits yellowing, leaf distortion, and severe mosaic. The papaya fruit will have bumps and a "ringspot".

PARTICULATE MATTER: pieces of matter small enough to remain suspended in the the air for long periods of time

PARTURITION: giving birth; synonymous with pupping

PATHOGEN: a bacterial, viral, or fungal agent of disease

PESTICIDE: a substance that is used to prevent or kill animals or insects that damage plants of crops

PETROLEUM HYDROCARBONS: the primary molecular constituent in oil, gasoline, diesel, and other petroleum products; the simplest of organic compounds containing only carbon and hydrogen atoms

PHLOEM: a complex tissue in the vascular system of high plants like trees; responsible for the translocation and storage of nutrients and water

PHOTOSYNTHESIS: the conversion of sunlight, carbon dioxide, and water into glucose and oxygen by organisms for use as energy

PLANKTIVOROUS: referring to planktivore, an aquatic organism that feeds on zooplankton and phytoplankton

PLANKTON: microscopic organisms that float freely in marine and fresh bodies of water PM10 particulate matter less than 10 micrometers in diameter

PM10: particulate matter less than 10 micrometers in diameter

POACHING: the illegal hunting, killing, or capturing of wild animals

POLLINATE: the transfer of pollen from the anther to the stigma of a plant often from one plan to another

POLYCHLORINATED BIPHENYLS (PCBS): a broad range of man-made chemicals used as a coolant and flame retardant; a chemical known to be carcinogenic, neurotoxic, and cause developmental and reproductive disorders

POLYMER SHEET: a chemical compound or mixture of compounds formed by polymerization and consisting essentially of repeating structural units

POLYMERASE CHAIN REACTION (PCR): a method to amplify copies of DNA using many cycles of DNA denaturation, primer annealing, and DNA polymerization

POLYP: the individual base of a coral; sedentary form of a coelenterate

POPULATION: a group of organisms that are of the same species, which reside in the same general geographic location and are able to interact and breed

PRECIPITATION: weather activity in the form of rainfall, hail, or snowfall

PRESERVE: a protected area set aside for the refuge and safety of plants and animals

PRIMARY FOREST: mature forests of native trees that have not been affected by deforestation

PROPAGANDA: use of communications to try and influence a person or group of people with often biased information

QUANTIFIABLE: the ability to be expressed or measured in terms of quantity

QUARANTINE: a state of enforced isolation, often imposed when a human or animal has been exposed to an infectious disease

RADIOACTIVE: atoms that spontaneously emit subatomic particles and/or energy; emitting ionizing radiation when decaying

RADIONUCLIDE: atoms with an unstable nucleus, characterized by the availability of excessive energy; can be naturally occurring or artificially produced

RANCH HAND: an U.S. Military Operation during the Vietnam War that involved spraying millions of liters of herbicide in order to rid Vietnam of crops and vegetation to expose roads and trails used by the Viet Cong

REDUCING EMISSIONS FROM DEFORESTATION AND FOREST EGRADATION (REDD): a mechanism to mitigate climate change through reducing net greenhouse gas emissions; under negotiation by the United Nations

RESERVOIR: a natural or artificial body of water that is used to store water

REVERSE OSMOSIS: process by which a solvent passes through a porous membrane in the direction opposite to that for natural osmosis when subjected to a hydrostatic pressure greater than the osmotic pressure

RNA: ribonucleic acid; a molecular structure present in all living cells that carries and transmits commands to build proteins

ROTOR: the non-stationary part of a rotary motor or generator; contains electromagnets which are used by the generator to induce an electrical current

ROUNDTABLE ON SUSTAINABLE PALM OIL (RSPO): a non-profit organization that aims to transform markets towards sustainable palm oil

RUNOFF: the portion of precipitation on land that is not absorbed by the ground but instead flows into nearby waterways; waste products that are carried by rain into surface waters

RUST BELT: referring to economic decline in a region of the United States surrounding the Great Lakes

SAFE DRINKING WATER ACT: Federal law setting standards for water quality across the United States to be enforced primarily by the U.S. Environmental Protection Agency

SALMONIDS: a fish belonging to the family Salmonidae which includes the salmon, trout, and whitefish

SANCTUARY: a protected area set aside for the refuge and safety of plants and animals

SECONDARY PRODUCTION: the rate at which primary consumer organisms convert food into their own biomass

SEDIMENT: a solid material that moves and settles in a different location, often from erosion

SESSILE: an organism that is immobile or fixed to one place

SHOAL: a piece of land where moving water promotes sediment deposition, such as a sandbar

SIERRA NEVADA MOUNTAIN RANGE: mountain range in the western United States between the Central Valley of California and the Basin and Range Province, the majority of the range lies in California

SNOWPACK: a layer of compacted snow on the ground that often accumulates in high altitudes for long periods of time

STAND: a contiguous area that contains a number of trees that are relatively homogeneous or have a common set of characteristics.

STATOR: the stationary part of a rotary motor or generator; contains conductors on which electric current is induced by the rotation of the rotor

SYMBIOTIC RELATIONSHIP: a close relationship between two organisms in which they mutually benefit

THERMAL PROCESS: a process by which phosphate rock is heated to create phosphorus pentoxide, which is then dissolved in dilute phosphoric acid to form very pure phosphoric acid

THYROID CANCER: cancer originating in the thyroid gland, which is located in the front of the neck

TOPOGRAPHY: surface features or an area or a region on a map

TRANSLOCATION: immobilizing and transporting one or more individuals of a population from an area of high population density to an area of lower population density where they will have less of an environmental impact

TROPICAL STORM: a localized, powerful storm that forms in tropical areas

TSUNAMI: a series of high sea waves caused by an earthquake, submarine landslide, or other disturbance

TURBIDITY: a measure of the amount of suspended solids in water

VASCULAR: of or relating to a channel for the conveyance of a body fluid (as blood of an animal or sap of a plant)

VIRUS: an acellular particle containing a genome that can replicate only inside a cell; a noncellular pathogen

WATER STRESS: a lack of sufficient water resources available to meet the demands of water usage

WATER TABLE: the planar level at which water saturated soil or rock meets unsaturated ground

WEANING: a period in which young mammals transition to sources of food other than their mother's milk

WESTERLIES: consisting of the winds that blow from the west towards the east; a belt of major air currents in the mid-latitudes of the northern and southern hemispheres

WET ACID PROCESS: a process by which phosphate rock is dissolved in sulfuric acid forming phosphoric acid and waste products

WICKET GATE: doors used to control the rate of water intake through a generator's turbine

WORLD HEALTH ORGANIZATION (WHO): an agency of the United Nations that promotes health and control of diseases around the world

REFERENCE

Bora, S., Ceccacci, I., Delgado, C. & Townsend, R. (2011). Food security and conflict. World Bank, Washington, DC. © World Bank. Retrieved from https://openknowledge.worldbank.org/handle/10986/11719. Available under Creative Commons Attribution License 3.0 (CC BY 3.0). Modified from original.

CK12. (2014). Photosynthesis. Retrieved from http://www.ck12.org/section/Photosynthesis-%3A%3Aof%3A%3A-MS-Cell-Functions-%3A%3Aof%3A%3A-CK-12-Life-Science-For-Middle-School/. Available under Creative Commons Attribution-NonCommercial 3.0 Unported License. (CC BY-NC 3.0). Modified from Original.

CK12. (2014). Photosynthesis. Retrieved from http://www.ck12.org/section/Photosynthesis-%3A%3Aof%3A%3A-MS-Cell-Functions-%3A%3Aof%3A%3A-CK-12-Life-Science-For-Middle-School/. Available under Creative Commons Attribution-NonCommercial 3.0 Unported License. (CC BY-NC 3.0). Modified from Original.

CK12. (2015). Food and nutrients. Accessed August 31, 2015 at http://www.ck12.org/user:a3F1aWNrQHdlYmIub3Jn/section/Food-and-Nutrients/. Available under Creative Commons Attribution-NonCommercial 3.0 Unported License. (CC BY-NC 3.0). Modified from original.

EEA. (1997). Towards sustainable development for local authorities – approaches, experiences and sources. Retrieved from http://www.eea.europa.eu/publications/GH-07-97-191-EN-C. Available under Creative Commons Attribution License 3.0 (CC BY 3.0). Modified from Original.

EEA. (1997). Towards sustainable development for local authorities – approaches, experiences and sources. Retrieved from http://www.eea.europa.eu/publications/GH-07-97-191-EN-C. Available under Creative Commons Attribution License 3.0 (CC BY 3.0). Modified from Original.

EEA. (2013). Combined heat and power. Retrieved from http://www.eea.europa.eu/data-and-maps/indic ators/combined-heat-and-power-chp-1. Available under Creative Commons Attribution License 3.0 (CC BY 3.0). Modified from original.

EPA. (n.d.). Climate change. Accessed August 31, 2015 at http://www.epa.gov/climatechange/. Modified from original.

Godheja, J. (2013). Impact of GMO'S on environment and human health. Recent Research In Science And Technology, 5(5). Retrieved from http://recent-science.com/index.php/rrst/article/view/17028. Available under Creative Commons Attribution License 3.0 (CC BY 3.0). Modified from original.

Kelly, L. (2005). The global integrated pest management facility. World Bank, Washington, DC. © World Bank. Retrieved from https://openknowledge.worldbank.org/handle/10986/19053. Available under Creative Commons Attribution License License 3.0 (CC BY 3.0). Modified from Original.

Kriebel, D., Tickner, J., Epstein, P., Lemons, J., Levins, R., Loechler, E. L., ... Stoto, M. (2001). The precautionary principle in environmental science. Environmental Health Perspectives, 109(9), 871–876. Retrieved from http://www.ncbi.nlm.nih.gov/pmc/articles/PMC1240435/.

Kriebel, D., Tickner, J., Epstein, P., Lemons, J., Levins, R., Loechler, E. L., ... Stoto, M. (2001). The precautionary principle in environmental science. Environmental Health Perspectives, 109(9), 871–876. Retrieved from http://www.ncbi.nlm.nih.gov/pmc/articles/PMC1240435/.

Maghari, B. M., & Ardekani, A. M. (2011). Genetically Modified Foods and Social Concerns. Avicenna Journal of Medical Biotechnology, 3(3), 109–117.

NAL. (2007). Sustainable agriculture: Definitions and terms. Retrieved from http://afsic.nal.usda.gov/sustainable-agriculture-definitions-and-terms-1#toc1. Modified from Original.

NSF. (2009). Transitions and tipping points in complex environmental systems. Retrieved September 24, 2015 from http://www.nsf.gov/geo/ere/ereweb/ac-ere/nsf6895_ere_report_090809.pdf. Modified from original.

NSF. (2009). Transitions and tipping points in complex environmental systems. Retrieved September 24, 2015 from http://www.nsf.gov/geo/ere/ereweb/ac-ere/nsf6895_ere_report_090809.pdf. Modified from original.

Nuckols, J. R., Ward, M. H., & Jarup, L. (2004). Using geographic information systems for exposure assessment in environmental epidemiology studies. Environmental Health Perspectives, 112(9), 1007–1015. doi:10.1289/ehp.6738.

Nuckols, J. R., Ward, M. H., & Jarup, L. (2004). Using geographic information systems for exposure assessment in environmental epidemiology studies. Environmental Health Perspectives, 112(9), 1007–1015. doi:10.1289/ehp.6738.

OpenStax College. (2013). Concepts of biology. Retrieved from http://cnx.org/contents/b3c1e1d2-839c-42b0-a314-e119a8aafbdd@8.10. OpenStax CNX. Available under Creative Commons Attribution License License 3.0 (CC BY 3.0). Modified from Original.

OpenStax College. (2013). Concepts of biology. Retrieved from http://cnx.org/contents/b3c1e1d2-839c-42b0-a314-e119a8aafbdd@8.10. OpenStax CNX. Available under Creative Commons Attribution License 3.0 (CC BY 3.0). Modified from Original.

OpenStax College. (2013). Concepts of biology. Retrieved from http://cnx.org/contents/b3c1e1d2-839c-42b0-a314-e119a8aafbdd@8.10. OpenStax CNX. Available under Creative Commons Attribution License License 3.0 (CC BY 3.0). Modified from Original.

OpenStax College. (2013). Concepts of biology. Retrieved from http://cnx.org/contents/b3c1e1d2-839c-42b0-a314-e119a8aafbdd@8.10. OpenStax CNX. Available under Creative Commons Attribution License 3.0 (CC BY 3.0). Modified from Original.

Page attribution: Essentials of Environmental Science by Kamala Doršner is licensed under CC BY 4.0. Modified from the original by Matthew R. Fisher. "Review Questions" is licensed under CC BY 4.0 by Matthew R. Fisher.

Page attribution: Essentials of Environmental Science by Kamala Doršner is licensed under CC BY 4.0. Modified from the original by Matthew R. Fisher. "Review Questions" is licensed under CC BY 4.0 by Matthew R. Fisher.

Page attribution: Essentials of Environmental Science by Kamala Doršner is licensed under CC BY 4.0. Modified from the original by Matthew R. Fisher. "Review Questions" is licensed under CC BY 4.0 by Matthew R. Fisher.

Page attribution: Essentials of Environmental Science by Kamala Doršner is licensed under CC BY 4.0. Modified from the original by Matthew R. Fisher. "Review Questions" is licensed under CC BY 4.0 by Matthew R. Fisher.

Theis T. & Tomkin J. (Eds.). (2015). Sustainability: A comprehensive foundation. OpenStax CNX. Retrieved from http://cnx.org/contents/1741effd-9cda-4b2b-a91e-003e6f587263@43.5. Available under Creative Commons Attribution License License 3.0 (CC BY 3.0). Modified from Original.

University of California College Prep. (2012). AP environmental science. Retrieved from http://cnx.org/content/col10548/1.2/. Available under Creative Commons Attribution 4.0 International License. (CC BY 4.0). Modified from original.

University of California College Prep. (2012). AP environmental science. Retrieved from http://cnx.org/content/col10548/1.2/. Available under Creative Commons Attribution 4.0 International License. (CC BY 4.0). Modified from original.

University of California College Prep. (2012). AP environmental science. Retrieved from http://cnx.org/content/col10548/1.2/. Available under Creative Commons Attribution 4.0 International License. (CC BY 4.0). Modified from original.

World Bank Group. (2015). Ending poverty and hunger by 2030: An agenda for the global food system. Washington, DC. © World Bank. Retrieved from https://openknowledge.worldbank.org/handle/10986/21771. Available under Creative Commons Attribution License 3.0 IGO (CC BY 3.0 IGO). Modified from original.

World Bank. (2004). Persistent organic pollutants: Backyards to borders. Washington, DC. © World Bank. Retrieved from https://openknowledge.worldbank.org/handle/10986/14896. Available under Creative Commons Attribution License License 3.0 (CC BY 3.0 IGO). Modified from Original.

World Bank; Food and Agriculture Organization; International Fund for Agricultural Development. (2009). Gender in agriculture sourcebook. Washington, DC: World Bank. © World Bank. Retrieved from https://openknowledge.worldbank.org/handle/10986/6603. Available under Creative Commons Attribution License License 3.0 (CC BY 3.0 IGO). Modified from Original.

INDEX